NORTH AMERICAN
WATERFOWL

North American Waterfowl

By

ALBERT M. DAY

DIRECTOR OF THE FISH AND WILDLIFE SERVICE
UNITED STATES DEPARTMENT OF THE INTERIOR

Illustrated

Sketches by Bob Hines

STACKPOLE AND HECK, Inc.

NEW YORK   1949   HARRISBURG

Copyright 1949
By
STACKPOLE AND HECK, INC.

All Rights Reserved

SK
361
D28

Printed in the United States of America
By The Telegraph Press, Harrisburg, Pennsylvania

TO

My young sons, Richard Thayer and John Kendall Day, with the fervent hope that they and their sons may also thrill to the fascination of the out-of-doors that I have known.

Preface

     MUSIC has left deep, intangible, and lasting impressions on the souls of men since the beginning of time. Every nation has developed melodies that suit the moods of its own people. Even in the deepest jungles native tribes follow the rituals of rhythm. Music has had much to do with shaping the lives and destinies of men, but of all music in human experience, I will still take the clarion call of the lead goose winging northward in spring migration as the sweetest sound of all. Those notes have real meaning. They signify that the long and dreary cold of winter is about done, that soon the snow banks will melt away and the green grass will again begin to color the drab, frozen landscape. They mean that the ancestral migration urge of the birds has been stirred with the departure of Old Man Winter, and that Spring is on its way. Such melodies of Nature are dear to the hearts of those who have lived close to her bosom.

    I shall never forget the first call of geese that came over our ranch in eastern Wyoming in the spring of 1914. We had gone through one of the toughest winters in recorded history. Blizzard after blizzard had roared out of the north with temperatures ranging from 20 to 40 degrees below zero. Snow was piled high in every direction. Our hay and grain were just about gone. We had seen our cattle and horses, weakened from the severe winter, waste away and perish before our very eyes. My Dad and I had fought our way through the mountainous snowbanks, wearily shoveling and breaking trail for the teams of horses that followed us, pulling sleds piled high with hay, struggling to reach the feed grounds where the cattle stood huddled together with rough winter coats caked with ice and snow. The pitiful look in the eyes of these starving animals left impressions that I will never forget.

After scattering the hay, we battled the blizzards back to the ranch house, usually finding it necessary to shovel the swirling and drifting snow away from the barn doors so that we could put our exhausted horses under shelter. I still recall how many times during that particular winter we tunneled through the drifts to the barn doors so we could get our livestock in and out for water.

Suddenly, one day, when it seemed as though the human body and mind could take no more of the punishing cold, the snow and wind that roared out of the northwest, the weather broke. One of those delightful, long-awaited "Chinooks" blew its warm breath out of the southwest. The hard surfaces of the snow started to soften. The tops of the fence posts began to emerge through the crust of that blinding, glistening wintry blanket. Little rivulets of water trickled out of the bases of the mountainous snowbanks.

And then, one wonderful night, I was awakened by sweet music from the sky. Winging up the valley from the south came a little band of Canada honkers calling out the most melodious notes that I have ever heard. The Chinook might not last long enough to take away all of the snow, but when the geese started north there was little doubt that spring would soon arrive to finish the job.

In the days when I was a youngster growing up on the plains of eastern Wyoming, wildlife was as common as cattle and saddle horses. Antelope were numerous and we thought it great sport to chase them with our ponies. I recall one early painful experience. My older brother, sister, and I rode to the little country school house, all of us on a long-barreled old bay mare that we knew as "Ginger." Coming home from school one evening, we put her into a dead run to see if we could catch up with a herd of antelope kicking up the dust as they bounded away from us across the prairie. As we raced after them, Ginger dropped a front foot into a badger hole. She and three youngsters sprawled in every direction. Being the youngest, my seat was nearest the rear of the old mare and as she went down I sailed head over heels through the Wyoming atmosphere, landing in a large bed of prickly pear. I still recall my mother's disgust as she pulled the cactus spines out of the rear end of her youngest son while he screamed to the high heavens.

Prairie dog towns extended for miles northward. Wise-looking little ground owls lived in the abandoned prairie dog holes, and rattlesnakes were always to be found in the near vicinity. Wolf packs ranged throughout this country in those days, and we could always expect a few cattle and an occasional colt to be killed to satisfy their ravenous appetites. My favorite pinto Indian saddle pony carried a large scar low on his left hind quarter where a wolf had slashed at him as he raced to escape the pack! The yip-yip-yip-e-e-e of the coyotes about the ranch house provided constant evening and early morning music. Sage hens were common. We youngsters, on several occasions, took eggs from the nests and brought them to the ranch where we hatched them out under Mother's setting hens. Invariably the experiment was a failure because the downy little fellows disappeared as soon as they were strong enough to scurry away into the surrounding grass and sage brush.

Here, I killed my first goose. We had some fine irrigated meadowland, and occasionally a flock of geese would drop in during the fall flight to feed on the tender shoots of alfalfa or the waste grain in the wheat and barley stubble. They always lit well out in the open and peered around intently for several minutes before starting to feed, and then the old leader always kept a sharp eye and ear out for danger.

One day—more years ago than I like to remember—I had been hunting ducks along the creek, jumping mallards out of the willow slough, when I heard a bunch of geese coming in far to the north. When they approached the open fields of the ranch, they started circling for a landing in the alfalfa patch. Losing all interest in ducks for the time being, I sneaked along the irrigation ditch for half a mile, screened by the willow-covered embankment. This cover bordered the east side of the field, and we usually had west winds, so I managed to get opposite them without being detected. I was still completely out of range for my dad's old 1897 Winchester pump, but closer to wild geese than I had ever been before.

There they stood, necks and heads craned high to make sure that all was well. I could hear them chattering while my heart pounded so loudly that I was sure its echo was what they were talking about. Finally, as they began to settle down, the old boss gander spotted a coyote trotting across the field toward them. With a loud call to the others he ran along the ground and took off, the others forming in behind him. And then my heart almost stopped beating. They were coming directly toward me! When their heavy wings swept overhead I raised the old gun and let fly—a goose dropped into the willows nearby. I can't tell to this day whether it was the one I aimed at or not, or whether I even aimed. But I had my first goose, and a coyote had helped me get it.

Mallards nested along the borders of the creeks and always afforded fine jump shooting in the fall, but our best duck hunting was over on the neighboring ranch on a small lake that had been constructed by building an earthen dam across a small running creek. The dam had grown over with willows and tall weeds. In the fall the water was literally covered with ducks that rested a bit as they moved southward in their fall migrations. Those birds really tasted good to growing youngsters whose appetites were never satisfied and whose diets contained little variety from the beefsteak and fried potato combination that now seems the last word in fine food.

Little did we think in those days that the time would soon come when this abundance of wildlife would be so drastically reduced!

Joining the U. S. Biological Survey in 1918, while still attending the University of Wyoming, one of my first tasks was to assist in a state-wide survey of the antelope. The best estimate that we could make at that time was that there were less than 7,000 of these magnificent animals in the entire state of Wyoming. They were so few in numbers that many people felt that the antelope would soon be added to that ever-growing list of vanished species. Fortunately, an awakened public consciousness gave them the protection they needed in time and they are once again abundant in many parts of that state.

In later years I have had the good fortune to associate with outdoor friends and hunters in many parts of the country. Professionally, my work has thrown me with game management workers, game breeders, sincere sportsmen of the highest type, ornithologists, protectionists, killers who would shoot without restraint, violators haled before the bar of justice, high officials of the Canadian and Mexican Governments, state and Federal scientists and administrators, and Congressmen and Senators who are themselves interested in wildlife problems or whose constituents either want something or are peeved about the state or Federal restrictions on their favorite outdoor sport.

Since the U. S. Fish and Wildlife Service must regulate the seasons on migratory waterfowl, assuming responsibilities imposed under treaties with Great Britain and the United Mexican States, it has fallen to my lot to participate in many public meetings to discuss the status of the nation's waterfowl and to sample the thinking of that vast group of outdoor enthusiasts which we know as the American duck hunter. Questions that arise in these public meetings as well as those that reach us through correspondence with the thousands of duck hunters throughout the country each year reveal, I am reluctantly convinced, a rather general lack of understanding of the basic principles of waterfowl management. Many are interested only in bag limits and seasons—the limitations on their own hunting privileges. Too few appear to give any thought to the fundamentals of production and protection, to the need for doing something to aid the birds instead of just satisfying their own desires.

In retrospect I am convinced that the vast bulk of the wildfowlers in this country want to make certain that the resource is so managed that "Young Sprout" may experience the thrill of whistling wings as he sits in a duck blind with the "Old Man" and sucks in his breath as he pulls down on that first mallard. They demand that waterfowl never be permitted to follow the sorry trail of greed and waste that took the passenger pigeon to extinction and the buffalo to the brink of oblivion.

Yet there are others, usually in small, noisy minorities, who have learned nothing from the pages of history. If unrestrained by some governing hand, they would have our ducks and geese wiped from the map in a few short years. Some frankly admit that duck hunting is an expensive sport and insist that the regulations should be kept liberal for the few who can afford it—"to heck with the Little Fellow!" Some can see nothing in waterfowl management but the annual regulations—the control of the kill—and they rant and rave if these are restrictive to the point of curbing their own sport. They give no thought to the obvious fact that waterfowl need management—that ducks and geese must have the right kind of environment in which to rear their young, sanctuaries where they may stop and feed unmolested on their semi-annual journeys each year, and suitable places to spend the winter months before they seek their mates and again wing northward to the nesting marshes.

After holding public meetings to discuss waterfowl management problems with duck hunters, administrators, conservationists, outdoor writers, and all who wanted to attend—in cities and towns in all parts of the United States—I can reach one conclusion rather clearly; that is, a conviction that there is real need for a better understanding of the basic principles of management. It is with this thought in mind that I have determined to attempt to pull together in one volume, in one place, if you please, the story of the yearlong studies and observations that precede the issuance of the annual regulations, the expensive and painstaking effort that goes into the purchase, improvement, and management of the wholly inadequate system of waterfowl refuges; to emphasize that we are still losing ground in protecting and in restoring habitat; that enforcement is wholly inadequate; that too many birds are still illegally taken during both the open and closed seasons; that hunting pressure has increased beyond all predictions, and that the over-all picture is far from bright. My sincere hope is that the pages that follow may bring this vital message home to the American public and will encourage greater effort to provide the kind of management this great resource so richly deserves.

THE AUTHOR

Acknowledgments

I AM INDEBTED to the Dominion Wildlife Service, Lands and Development Services Branch, Department of Mines and Resources, Canada, for preparing the chapter on "Waterfowl Conservation in Canada." I am grateful to Senor Luis Macias for reviewing the chapter on "The Mexican Waterfowl Situation." I also wish to express sincere appreciation to Dr. Clarence Cottam, Frederick C. Lincoln, Robert M. Rutherford, Dr. Ira N. Gabrielson, Frank Dufresne, Edna N. Sater, Dr. Logan Bennett, Dr. John W. Aldrich, J. Clark Salyer, Oscar H. Johnson, Jesse F. Thompson, and David R. Gascoyne, all of whom assisted in the critical review of portions of the manuscript.

I am most grateful to J. N. "Ding" Darling, my former Chief and friend of many years, the outstanding conservationist who, with his pointed cartoons and pungent writings, has brought so many folks to the realization of the importance of our natural resources, for his introduction to this book.

To the following publishers and authors, appreciation is due for permission to quote from copyrighted material:

    Doubleday and Co. Inc., for extracts from, "The Saga of Andy Burnett."

Mrs. John C. Phillips, for excerpts from, "Shooting Stands of Eastern Massachusetts."

Lester Velie, for quotes from his article in *Collier's*, May 15, 1948, "What Are We Going to Do for Water?"

Fairfield Osborn, President of the New York Zoological Society, for permission to quote from Dr. W. T. Hornaday's book, "Our Vanishing Wildlife."

Manley Miner for permission to quote from his father's book, "Jack Miner and the Birds."

Hugh Grey, editor of *Field and Stream*, for the use of portions of Harold Titus' article, "The Truth About the Izaak Walton League."

William Morrow and Co., Inc., for excerpts from, "Duck Shooting Along the Atlantic Tidewater," edited by Eugene V. Connett.

Dr. A. Brownley Hodges, of Norfolk, Va., for permission to copy pages from the Game Register of the Pocahontas Fowling Club.

Reilly & Lee Co., for permission to use "For Fish and Birds," from "Collected Verse of Edgar A. Guest."

Also to Robert (Bob) Hines, for his realistic and lifelike sketches of waterfowl, I wish to express my gratitude. He spent many long evenings and Sundays at the task, when instead he might have been enjoying himself more by dropping a fly or plug into some quiet pool.

Last, but not by any means the least, I am indebted to my wife, Eva Kendall Day, whose counsel helped to clarify and increase the interest of many portions of the manuscript, and whose untiring efforts in preparing it made possible the completion of this book, written as it was during the press of a multitude of responsibilities.

All photographs are from the official files of the U. S. Fish and Wildlife Service unless otherwise noted.

A. M. D.

Washington, D. C.
July, 1949

Layout and design by George F. Miley

Table of Contents

	INTRODUCTION—by J. N. "Ding" Darling	XIX
I	SKETCHES OF EARLY AMERICA - - -	1
II	THE AFTERMATH - - - - - - - -	21
III	THE SLOW AWAKENING - - - - -	31
	Wild Things in Relation to Agriculture	
IV	PROTECTIVE LEGISLATION - - - - -	39
	The Lacey Act	
	The Weeks-McLean Law	
	The Migratory Bird Treaty Act	
	The Act is as Solid as a Rock	
V	BIRD MIGRATION - - - - - - - -	55
	Bird Banding	
	Unusual Records	
	Practical Results of Banding	
VI	THE FLYWAY CONCEPT - - - - - -	69
	The Atlantic Flyway	
	The Mississippi Flyway	
	The Central Flyway	
	The Pacific Flyway	
	Management by Flyways	

VII	KEEPING UP WITH THE WATERFOWL	89
	Transect Sampling	
	Conditions on the Nesting Grounds	
	Winter Inventory	
	The Take by Hunters	
VIII	THE REGULATIONS - - - - - - - -	105
	How the Regulations are Made	
IX	WILDLIFE "G-MEN" - - - - - - - - -	115
	The Lacey Act	
	Violators Take Heavy Toll	
X	THE SANCTUARY IDEA - - - - -	145
XI	THE REFUGE SYSTEM - - - - - -	155
	Breeding Refuges—Upper and Lower Souris and Des Lacs Refuges in North Dakota	
	Malheur National Wildlife Refuge	
	California Waterfowl Refuges	
	Mattamuskeet and Adjacent National Wildlife Refuges	
	Wildlife Refuges Protect Endangered Species	
	Other Refuges	
XII	REFUGE DEVELOPMENT AND MANAGEMENT - - - - - - - - - - - - -	181
	Grazing	
	Controlled Burning	
	Timber	
	Farming	
	Muskrats	
	Beavers	
	Predators	
	Unusual Products	
	Recreation	
	Financial Returns to Counties	
	Upland Game	
	Trapping and Transplanting	
	Establishing New Breeding Colonies	

TABLE OF CONTENTS　　　　　　　xvii

    Public Hunting
    Field Trials
    Refuges are Community Assets

XIII RESEARCH INTO PROBLEMS OF WATERFOWL MANAGEMENT - - - - - - - - - 201
    Cooperative Units
    Lead Poisoning
    Botulism
    Predation
    Depredations
    Food Habits
    Pest Plant Control
    Insecticides
    Cripple Losses
    Many Questions Still Unanswered

XIV STATE ACTIVITIES - - - - - - - - 219
    Early State Interest in Waterfowl
    The Depression Thirties

XV THE PITTMAN-ROBERTSON PROGRAM 229
    Waterfowl Surveys and Investigations

XVI WATERFOWL CONSERVATION IN CANADA - - - - - - - - - - - - - - 249
    Early Federal Legislation
    The Migratory Birds Convention Act
    Federal Conservation Machinery
    Migratory Bird Sanctuaries
    Provincial Conservation Measures
    Private Conservation Activities
    International Cooperation

XVII THE MEXICAN WATERFOWL SITUATION - - - - - - - - - - - - - - 265
    Forestal y de Caza
    The Valley of Mexico
    Lake Tultengo
    Lake Texcoco
    Mexican Waterfowl Kill

XVIII	CONSERVATION ORGANIZATIONS	289
	The National Audubon Society	
	The Wildlife Management Institute	
	The Izaak Walton League of America	
	Ducks Unlimited	
	National Wildlife Federation	
	Outdoor Writers Association of America	
	Other Organizations	
	Individuals and Clubs	
XIX	YOUR WATERFOWL	315
	BIBLIOGRAPHY	321

Introduction

IF I HAVE ever read an honest book based on a thorough knowledge of the subject of wild duck management, this is it. Every duck hunter should read it, and if I know duck hunters they will eat it up and be grateful to you for your convincing frankness and courage.

Having been in your shoes during the duck crisis of 1934 to '36, I can understand why you felt the urge to write this book. Years gone by I had been one of those one-gallus duck hunters who looked upon the gol-darned Bureaucrats down in Washington with about the same skepticism that a Tennessee "Mountain Dew" maker looks upon a "Revenooer." When I went to Washington to head up the Duck Restoration Program, in the words of Mark Anthony, I went "to bury Caesar, not to praise him."

But those responsible for making the duck hunting regulations, whom I had thought a lot of badly informed sentimentalists, turned out to have among them not only ardent duck hunters and remarkable wing-shots, but the best-informed group of men I had ever known. Competent and devoted as they were, almost to the point of martyrdom, they were nevertheless hamstrung and chained in their cells by the bonds of badly-informed Congressmen who in turn were carrying out the demands of the equally misinformed duck hunting constituents who fed on fairy tales of rafts of non-existing ducks on flyways other than their own.

You, Al, were among that group when I went to Washington as Chief of the old Biological Survey. You cannot boast about yourself nor about the members of your staff who have battled Congress, battled Secretaries, battled jealous state executives, and battled the duck hunters themselves in the effort to keep enough ducks and enough marshes left to save the great American sport. You can't boast, but I can—now that I am long removed from the organization and sit and watch from the sidelines. And if Remington Arms, Winchester, Browning, Stevens, and duPont knew of the sacrifices and efforts contributed by such men as you, Clarence Cottam, your Assistant Director, J. C. Salyer, in charge of the Refuge System, and a long list of field men, to preserve the time-honored sport of duck hunting for the U.S.A., they would set up an endowment for all of you for life.

And sitting up nights after the day's work is done to turn out such a book as this is only one small evidence of that kind of zeal, courage, and devotion. It is "distinguished service beyond the call of duty."

It is the best book on ducks I have ever seen. Every duck hunter should read it.

May 14, 1949

Sincerely,
Ding Darling.

CHAPTER I

Sketches of Early America

*D*URING one of the public waterfowl hearings on the West coast, a stately-appearing old gentleman arose and with considerable eloquence and great vehemence, stated:

"I represent the Pioneer Association, the oldest organization in this state. My father was one of the first children born in this territory and I have lived here all of my life. I know from my own experience and the stories that have come to me from many of the old pioneers, that something is radically wrong with the present system of managing waterfowl. When the Federal Fish and Wildlife Service cut the season to 30 days, and the daily bag limit to four ducks, the lowest in the nation's history, they admitted their inability to manage migratory waterfowl. When we could use feed and have live decoys there were plenty of ducks in this area. The birds were fat, they remained for a long period. Since feeding and live decoys were prohibited the duck life has steadily decreased. I condemn the Fish and Wildlife Service for prohibiting these fine sporting practices. They have completely ruined our duck hunting."

Many of the old pioneers have laid their guns aside, content to talk about the glorious experiences of their past, but others, typified by this disgruntled old gentleman, cause the Federal and state conservation agencies no end of trouble. They still live in the days of

Plate 1. Great flocks of ducks and geese greeted Andy Burnett as he rode down into California's Central Valley in 1832.

their youth—days when waterfowl were so abundant that they could kill any number by any method, days when there were abundant marshlands instead of the towns and highways and airports and agricultural sections with fields of potatoes, corn, and rice. All of these now symbolize America. These gentlemen shudder to think of being restricted to a miserly bag limit of four or five ducks a day as their thoughts revert to that squally Friday during the fall of 1902 when they filled the duck boat without any trouble at all and gave birds away to practically everybody in town. Those were really the good old days.

Wildfowling would not have survived this long if that kind of hunting had been permitted to continue. Fortunately, restraints were imposed in time to save the sport for future generations. At least it is the hope of all true conservationists that adequate safeguards have now been established so that duck and goose hunting may always be an American institution. To gain a better understanding of the thinking of some of these "old timers," it is worth while to review bits of early American literature. These will establish the background for the psychology that has influenced and often hampered adequate management of the resource in our modern times.

For a recital of early game conditions and of waterfowl abundance on the West coast, let us turn to Stewart Edward White's delightful historical novel, "The Saga of Andy Burnett." It is a fascinating story of an adventurous young Pennsylvania lad whose constant companion was the Long Rifle originally owned by Daniel Boone and handed down to Andy from his grandfather, who had fought Indians with Boone in Kentucky. The book carries him through the early fur-trapping days of the West, and finally into California where he settled for the balance of his days. Here is a short quotation:

Andy Burnett reached California in the autumn of 1832. It is impossible to say exactly where he and his three horses crossed the Sierras. He may have found his way past what is now called Donner Lake and so over the summit and down the American River; or, more probably, he may have picked a route farther north. * * * *

That night he slept among the cottonwoods. Next day he rode through the waist-high grasses. Soon he ran into swarms of game—elk, antelope, deer in herds throughout the bottoms. Andy had never seen so many waterfowl; had never imagined there could be so many, anywhere. They covered the surface of the small lakes so thickly that, even from the vantage of his saddle, Andy could discern but a gleam of water here and there. As he rode nearer he found the borders of these ponds eaten, or beaten, bare for a hundred feet or more to form muddy beaches, and that the beaches too were covered with the ducks and geese, basking or preening in the sun. He rode among them. They waddled away, to right or left, shaking their

Plate 2. "Mallards Coming in for a Landing." The first duck stamp. Drawn by J. N. "Ding" Darling, noted cartoonist and one-time Chief of the Bureau of Biological Survey. This stamp was issued in 1934. Sale 635,001.

tails in indignation, gabbling their opinion of discourtesy. But they did not fly. Andy waved his arms and shouted. A few of the more nervous sprang into the air; but immediately dropped down again. But every head was turned accusingly toward the intruder; and into the peaceful air arose a mighty chorus of quackings and honkings.

"Now look here!" he expostulated, speaking aloud after the fashion of the man who spends much of his year in solitude, "I've got as much right here as you have! I never in my life been so cussed out!"

A goose stretched his neck at him, hissing, seeming almost decided to nip a horse's legs. Andy turned his smile toward this valiant but somewhat ridiculous champion.

"No, you don't, old girl," said he. "I've got a trick of my own!"

On a sudden impulse he extended the long rifle and fired it into the air.

A blank instant silence followed the sharp crack of the piece, broken a half second later by the crash as of a mighty waterfall as the birds took wing. It seemed as if the dark earth were lifting to expose the hidden silver of the lake. The air was filled with hurtling bodies. The very sky was darkened. And another great roar, and a third, like successive peals of thunder, rolled across the man's astonishment; and then a smooth high silence made up of the thin whistlings of thousands upon thousands of wings. But the startlement was only momentary. Almost at once the birds began to drop back to their resting places, at first by ones and twos, then by tens and twenties, finally by thousands. Each volplaned down at a sharp angle, upended, lowered yellow webbed feet, set the brakes of outspread pinions, furrowed the water in a long silver curve, and came to rest, wagging his tail vigorously, fluffing his plumes, and quack-quacking his resentment at the shock to his nerves. Andy threw back his head and laughed joyously.

Such was California in the early days. That state is still a great concentration point for wintering waterfowl—more so, perhaps, than any part of the United States. For this, West coast duck hunters can be thankful, although they do not enjoy all of the same kinds of sport that are found elsewhere. Nature seems to have deliberately apportioned her bounties among the different sections of the country. Some furnish habitat for antelope, some for elk, or deer, some for quail; others for woodcock or grouse or turkeys. Coastal waters provide abundant sport and food fishes while inland streams and lakes produce trout, bass, pike, muskellunge or catfish. Seldom do the natural conditions in any one area combine to provide the environmental requirements for more than a few of the numerous species which the American public has learned to take for sport.

New England is a good example. Nowhere else is there such diversity and abundance of sport and food fishes. Delicious clams are found on the tidal beaches, with oysters and lobsters in the coastal waters. Striped bass abound, while record-breaking tuna and bluefish provide some of the finest sport fishing in the world. Gloucester and Boston are the seafood centers where cod, haddock, pollock, mackerel and other commercial species from the North Atlantic are unloaded. New England ranks tops in the sport and commercial fishing field.

But with waterfowl, the story is different. The tree-bordered lakes and sparse marshes scattered among the rock-ribbed hills of the northeastern United States provide good breeding grounds for woodcocks and black ducks, and these are about all. The coastal bays furnish food for rafts of bluebills and other divers during the fall migration, but there is little to attract geese. Normally, they move southward to their ancestral wintering grounds of the Maryland, Virginia, and Carolina coasts with little more than a chattering chorus of honks and calls as they pass over New England's stony hills and rocky coast line.

But the honkers did not always fly over eastern Massachusetts without stopping. Prior to 1935 when the Federal regulations outlawed the use of bait and live decoys, geese were lured out of the skies by trained tame callers and brought to blinds which had been well "conditioned" with wagon-loads of grain scattered in front of them.

John C. Phillips, a noted ornithologist and sportsman of Massachusetts, in his book entitled, "Shooting-Stands of Eastern Massachusetts," published in 1929, describes the shooting on Silver Lake, six miles west of Kingston Bay, as follows:

Here the decoying of migrating geese by the use of flyers first reached its perfection. At Arnold's, on the west shore, I have seen 50 or 60 goslings released at the same time, setting up a racket of goose music impossible to realize unless one has been there. Forty or fifty years ago a stand of 20 geese was considered large, but when this stand (Arnold's) was in its heyday between 300 and 400 were in use!

Describing "Long Point on the east shore in Pembroke," Mr. Phillips tells us:

In 1925 they shot 107 geese and 125 ducks, using live goose decoys and a large rig of live ducks and blocks. In 1926, 76 geese and 60 ducks. This last was accounted a very poor season—"Not many geese seen. Plenty of duck in the Bay, but did not show up here."

When Farrar gunned this stand, on a year when Widgeon Point was closed, they got very nearly 400 geese. This was high line and was in 1920 or 1921.

Of "Reservoir Pond" consisting of 225 acres near Canton, Massachusetts, Dr. Phillips continues:

The Birch Point Stand has been in operation since 1898. The blind is about 75 feet long and they use 75 ducks and 30 live geese. Their best bag was 150 ducks and 125 geese. They average about 100 ducks and 75 geese. Best year, 1924. They once got 24 Black Ducks with one discharge from three guns.

Another bit of history I got from Lawrence Horton, of Canton, who gunned so long for Talbot Aldrich at Ponkapoag. He finds only two years' records dealing with his shooting at Reservoir Pond. In 1889, he shot 219 ducks and geese, and in 1890, 155. In 1891, Joe Revere tells him they shot

325 birds, a large bag. Three hundred and fifty was about high line for this stand and 50 geese about the average.

An interesting account of how the live goose decoys were handled is found in a chapter entitled "Cape Cod Memories," by A. Elmer Crowell, in Eugene V. Connett's recent book, "Duck Shooting along the Atlantic Tidewater." The period was around 1890. Excerpts are:

The next fall Mr. Charles Hardy and others had the Three Bear Camp built on Pico Point, and I had the care and handling of the decoys. I bought a large number of live geese from a friend in Hanover, Massachusetts. It made, with what we had, about fifty in all. This blind was about half a mile from mine to the west. We had a large beach made out into the lake and pinned twenty geese on it. Then we built pens on the hill back of the blind, attached electric wires to them, and ran the wires down to the blind. We had four boxes on the inside of the blind with electric buttons.

When we saw a bunch of geese coming we pushed the buttons and the trap door would fall down and out would come the flyers. They would fly out over the lake, and sometimes they would join the bunch of wild ones and bring them to beach. It was a success and we had good shooting for a number of years. At that time we began to bait the small ponds with corn, and two years later we baited in front of our blinds at the lake. It stopped the Black Ducks from going south, so the ponds were full of ducks; the shooting was great. But we could not sell them in the markets, as the law cut it out. Soon the law cut out the live decoys, and that was the end of good shooting here.

Mr. Crowell ruefully ends his chapter with this comment:

I am eighty-four years old and my mind wanders back to the good old days when there was no law on birds. They were the days for me!

There are still a few survivors of that era who have persistently and vociferously fought for the return of bait and live decoys for goose hunting in Eastern Massachusetts so they might again have the kind of shooting they enjoyed in those "good old days." It would be safe to give odds that they will continue to be disappointed.

Duck hunting along the New England coast in early days was excellent as the black ducks gathered in the bays and marshes in the fall, and the scaups and ruddies and sometimes the redheads stopped over to feed during the long flights between their nesting grounds and winter quarters. Again we quote from John C. Phillips' "Shooting Stands of Eastern Massachusetts":

Mr. "Cop" (Lawrence) Horton, of Ponkapoag, who has gunned all his life at Reservoir Pond and Ponkapoag, tells me that forty years ago "Brad" Cobb and C. N. Draper gunned these Neponset Meadows, probably the same stand I remember about 1897. He described it as below the "Horseshoe." He does not know how many ducks they used to shoot there, but he recalls one terrific story that Cobb told. Cobb said that one day they shot so many ducks he had to go home and get his horse and tip-cart to carry them home. They were heaped so high on the cart that Draper walked behind the cart to pick up those that joggled off!

Also:

Much of the sport was in seeing how many sitting birds could be killed with one shot—a practice long since outlawed as unsportsmanlike. Read the record of the Wenham Lake stands:

It is always interesting to remember the big days and big shots at Wenham, especially as these occasions were never frequent. I have a record of every big shot made so that I can safely give the best ones. One gun means one man with a double-barrelled gun, usually 10 ga. or 8 ga.; in later years 12 ga.

October 9, 1900. One gun; 8 Scaup out of 10 Scaup.
November 16, 1900. Two guns; 14 Blacks out of 19 (lost one).
October 22, 1901. One gun (J.C.P.): 8 Redheads out of 11.
October 25, 1901. Three guns; 11 Redheads out of 15 (two 8 ga. and one 10 ga. gun).
November 24, 1901. Three guns; 14 geese out of large flock.
October 8, 1903. Three guns; 13 Black Ducks out of 24.
October 22, 1903. Three guns; 16 Redheads out of 16 (or possibly 18). This was the best shot ever made at Wenham.
October 22, 1904. One gun (E. Crowell); 12 Scaups out of 18.
October 4, 1905. One gun (E. Crowell); 9 Ruddies out of 22.
October 8, 1914. One gun; 9 geese out of 14.
October 11, 1916. Three guns; 11 Black Ducks out of 15.
October 21, 1922. One gun (J.C.P.); 9 Redheads out of 10.
October 26, 1928. One gun (P. J. Perkins); 9 Black Ducks out of 15.

As to the "good seasons," we learn this:

Suntaug Lake had two stands in the old days, and a man named Grimson shot with George Williams. He says their best year was 240 large ducks (Blacks and Mallards), besides birds which were not counted. The average number taken at the stand would be about 100 to 125 good ducks. Speaking of big shots, Williams told me that one morning he and Grimson stopped 20 Black Ducks at one shot and picked up 19 of them, but a skunk got the twentieth. This was the biggest shot he ever made in his entire shooting experience. One moonlight night he got 13 Black Ducks at one shot alone. He used two muzzle-loading guns; one an 8-gauge single gun, and the other a double 11-gauge.

Eugene V. Connett, writing of the early shooting on Great South Bay of Massachusetts, states:

Lying on West Bog one day, a good many years ago with Will Corwin, he told me of a day's battery shooting which Captain Charles S. Hawkins had in 1878, gunning for market.

* * * *

On the day to which I refer, these two men returned to Bellport with six hundred and forty ducks, most Broadbill, which they shipped to New York at twenty-five cents a pair, F.O.B. Bellport! Captain Corwin "kept account of guns fired on the cabin slate, which was fastened to the sliding hatch, and always looked for and picked up plenty more dead ducks than he counted guns fired," to quote his son's letter.

A notable bag of birds was shot by Judge H. A. Bergen on Titus Bog * * *. This was in the days of the Bellport Gun Club. Guided by Richard B. Hamel, the Judge killed sixty-four Broadbill, ninety-eight Blackduck and one Gadwall between 9 a. m. and 1 p. m.! This record is taken from the Bellport Gun Club camp register and is signed by Judge Bergen. * * *

When the bag limit was twenty-five ducks a day, Wilbur states that many, many days they would take up the battery at eight, nine or ten o'clock in the morning, with the limit of fifty birds for the two gunners. When Wilbur's father and uncles, R. B. Hamel, John and Ed Corwin, gunned for market "the one-day bag to ship to market for one of them in an ice hole used to be ninety-seven Redheads," Wilbur writes me. "I do not mean that happened too often, but it happened to each of them. And when I was a boy John Petty did it once off in the ice hole northwest of Fiddleton Flat."

About 1897, I remember sailing with my parents from Bellport to Smith's Point in Barney King's big catboat. A flight of ducks went over that literally darkened the sun. I presume that they were Broadbill, and I hesitate to guess at how many tens of thousands of birds I was looking at.

A good idea of the lethal methods in early use along the Atlantic coast extending from Connecticut to the lower Chesapeake Bay can be obtained from excerpts of a chapter in, "Duck Shooting Along the Atlantic Tidewater," by Charles E. Wheeler. Describing the shooting along Connecticut's shore of Long Island Sound in the period around 1870, he writes:

The record kill for this area was made with a double-barreled, four-gauge gun owner by the writer, when eighty-one dead Broadbill were gathered and forty-six cripples were shot over with the twelve, making a total of one hundred and twenty-seven birds. Other kills with this gun of upward of fifty birds were not uncommon. Broadbill paid the heaviest toll, as they sat closely together in great flocks, thus offering an excellent target. Large kills of Black Ducks, Geese and Brant were also listed in the record of this "killer." * * * *

The battery was a much larger outfit and had to be carried in a large skiff or aboard a sail or power boat. It consisted of a box to accommodate one to four gunners, built in the center of a floating deck which, in the case of a one-man rig, would be about twelve feet long by six feet wide, with a folding canvas wing on each side and a canvas head wing across the bow. When all spread out it covered an area of about eighteen feet long by twelve feet wide. It was anchored on the shooting ground with bow and stern anchors, a large setting of decoys was anchored around it and sufficient lead or iron decoys were placed on the deck to settle it down to the level of the water. * * * *

The largest battery of record for this area was a four-gun rig, with a deck twenty-five feet square, upon which a man could walk. This was surrounded with three to four hundred floating decoys and another hundred were fastened to the deck at close intervals, making the entire outfit look like a veritable island of birds. It was set out in the bay just before opening day and left there for the entire season. The gunners lived aboard a cabin craft, which was the tender during the day and the hotel during the night. Thus they were on the job from daylight until dark every day. When shipments of fowl had to be made, the tender did the trip ashore at night. This rig was finally outlawed.

There were two types of bush blinds, one a floating device, rectangular in shape and large enough to conceal a skiff or duck boat. It was made of two by six plank with holes bored at close intervals to hold pea brush or cedar tops cut about three feet long. This blind was anchored at one end only, so it would swing with the wind. It was left on the feeding grounds throughout the open season. The surrounding area, within gunshot, was kept well baited so that the fowl became accustomed to the blind and fell an easy prey to the gunner with a good rig of decoys. The other type of bush blind was a staked blind. This, too, was rectangular in shape and of the same size as the "floater," the difference being that the brush was cut longer and stuck into the bottom of the bay instead of into a floating frame.

Still another method of hunting waterfowl practiced to quite some extent can hardly be called stalking, as that term implies sneaking up on the target, nor can it be called ambushing, as that suggests lying in wait for the game to approach. It might be called chasing, as it was done by running down wind to the birds in either a sail or power boat. It was generally known that ducks took wing against the wind, so it was quite a popular sport to sail up to windward of a flock of birds, swing off before the wind and sail directly toward the flock. * * * * As they flushed, the slaughter began, for by now the repeating shotguns were popular duck guns. The cannonading was terrific and the casualties equally high in killed and wounded. In both sail and power boat hunts the faster the boats

Plate 3. Who had the most fun? These two hunters who display perfectly legal takes back in the "good old days" of spring shooting in Missouri (1913) or the chap who killed his legal limit of 2 geese and 5 ducks in 1948—35 years later? (Photos by U. S. Forest Service and Fish and Wildlife Service.

and the more guns aboard, the greater the kill. Some hunts really were shameful. * * * *

And so the persistent hunt for wildfowl went on every day of the open season. Many thousands of fowl were slain—in fact, it seems almost a miracle that enough survived running the gauntlet from north to south and back from south to north to constitute breeding stock for another crop.

Two other excerpts taken from chapters by Frederick Barbour in the same volume portray the early shooting practices along the coast line of Virginia. They also reveal the sentiments of most of the "old timers" who now look back in retrospect and wonder how the hunters of those early days could possibly have been so wasteful.

The present-day gunner would be simply appalled at what was taken as a matter of course in the way of a day's bag in, say, the 1880's. I believe, on certain of the South Sea Islands, there is no concept of the word "weather," because one day is much like another. In the same way, the thought or idea of a bag limit never entered one's mind. I am sure it would be of interest to the reader to see some photostatic copies of the entries which appear in some of the old game books of the Princess Anne Club, Seelinger's marsh, and at Sand Bridge.

I recall an instance where sixty-eight Canvasbacks were gathered on the first pick-up when shooting from Shell Point on Long Island. The birds came in so fast that there was no lull in the shooting to make an earlier pick-up possible. No imputations of guilt were attached to such scores and such an episode was merely regarded as a grand day when conditions were right. I can recall my father killing sixty-three geese, when in the Teal Island blind situated on a property called Barbour's Hill.

Numbers in possession were unlimited and it should be noted that all these birds, of course, were shipped to various and sundry friends and so were not wasted. In retrospect, it seems hardly conceivable that a pair of Canada Geese should constitute a limit. In times past, the gunning season extended through February. During the winter months, the bays would freeze tight, with the exception of an airhole here and there. I recall a wildfowling confrere of mine considering himself quite fortunate in knocking off one hundred and six Ruddies in such an airhole.

Today, of course, such slaughter seems absolutely indefensible. But as I have said before, at the time these performances were chalked up they were regarded merely as outstanding days with thousands of birds available. The Raised Eyebrow Department was conspicuous by its absence.

And Richard L. Parks, writing of the period around 1890 on Virginia's Eastern Shore, says:

It took nearly a century of unmerciful slaughter to make a noticeable dent in the countless thousands of waterfowl that once frequented the

Eastern Shore, especially the Seaside marshes. But in the last half century their numbers have shrunk so fast that today it is scarcely worth while for even the most ardent wildfowler to maintain a boat, blinds, and decoys, while the clubs have faded out of existence and it is difficult even to engage a local guide equipped to furnish a decent day's sport.

Shooting alone has not brought the sport to its present low ebb. Broadly, of course, the decline has come as a result of man's throwing to the four winds the balance of nature. Drainage, drought, waste of timber, futile farming of submarginal lands—all have done their share. Such predators as fox, mink, wildcat, lynx, snakes, hawks, and owls have played a leading part. And man, the most voracious and bloodthirsty predator of them all, has not only killed the ducks, but so many other natural foods of the predators—the ones that seldom kill except to eat—that they have had to hunt still harder the small game he left. The while the "varmints" were left to increase by leaps and bounds.

Would that more of the old pioneers—the old timers—shared the sentiments of Fredrick Barbour and Richard Parks! All too many still demand that the hunting regulations be made liberal enough for them to spend their last hunting days in the shooting blind slaughtering ducks and geese as they did 50 years ago. Some accept progress in every other line yet refuse to recognize the changed conditions that ducks and geese must face in our modern economy.

A few of them even claim that there are as many birds as there were forty years ago. I suggest they review some of the old game logs as I was privileged to do with the register of the Pocahontas Fowling Club, Inc., which has operated continuously on Virginia's Back Bay since 1904. The following are the records from single pages recording the hunting success for comparable periods in 1904, 1918, 1934 and 1948. Note the progressive decline!

GAME REGISTER—Pocahontas Fowling Club, Inc.

1904 Date	Name	Black	Mallard	Sprig	Widgeon	Ruddy	Black head	Red head	Canvas	Misc.	Geese	Total
Dec. 19	Horace L. Hames	9	20	1	2	..	1	..	..	1	..	34
	M. Johnson	4	3	..	..	..	1	..	..	20	..	28
	A. Brooke Taylor	7	12	..	..	..	..	..	..	..	..	19
	Dick B. Taylor	6	13	..	..	..	..	..	1	..	..	20
20	Tazewell Taylor	4	..	1	1	7	3	..	3	6	1	25
	E. L. Mayer	4	2	..	6	7	4	..	..	13	1	30
	M. Johnson	3	2	1	..	..	1	..	..	3	1	11
	Horace L. Hames	2	2	..	..	1	..	..	2	..	..	7
	A. B. Taylor	..	..	..	2	..	..	1	1	..	1	5
	R. B. Taylor	..	..	..	1	2	1	1	..	..	1	6
22	Homer Swathing } S. D. Taylor	..	..	..	1	3	3	..	1	..	2	10
23	Frank Hitch	5	4	2	1	..	1	1	..	..	..	14
	Frank Smith Hitch	..	3	1	1	..	..	..	..	..	2	7
	Edet Bradford	3	5	1	..	..	1	..	1	2	2	14
26	P. S. Stephenson	..	..	4	7	1	1	..	1	..	..	14
	A. B. Taylor	..	..	1	1	..	2	3	..	..	..	7
27	Clarence Grinalds	3	2	15	3	3	3	..	..	5	..	34
	E. L. Mayer, Sr.	4	3	2	14	42	2	3	..	..	2	72
	C. F. Burroughs	..	3	..	1	12	..	2	..	..	2	20
	P. S. Stephenson	3	8	12	4	..	..	..	..	1	2	30
	A. B. Taylor	..	2	..	1	..	1	..	..	..	1	5
29	Frank Hitch	1	..	1	..	..	..	..	..	..	..	2
	Taylor 7	3	11	21	2	..	2	..	..	..	..	37
	D. L. Growner	..	..	..	..	4	..	1	1	10	..	19

Information from sample page, game log year 1904-1905. Season record begins September 30, 1904, closes March 20, 1905. Season total: Man-days hunting, 15½; total bag, 242½; average daily bag, 16.

1918 Date	Name	Black	Mallard	Sprig	Widgeon	Ruddy	Black head	Red head	Canvas	Misc.	Geese	Total
Dec. 9	John Paine	3	1	1	..	..	..	1	..	3	2	11
10	John Paine	2	..	2	1	..	1	..	..	1	9	16
	Edith Foote	..	..	..	..	..	..	..	..	..	..	..
	C. Hathaway	1	..	..	..	..	2	..	..	..	..	3
12	John Payne	1	3	..	..	..	..	..	..	2	..	8
	Chas. Hathaway	5	4	..	..	..	..	2	..	..	..	9
	Edw. Foote	1	3	..	..	1	1	2	..	1	4	13
13	John Paine	..	1	..	..	..	..	..	..	..	..	2
	Chas. Hathaway	1	..	..	1	..	..	1	1	1	..	5
	Edw. Foote	2	..	..	..	..	..	1	1	..	2	7
16	Francis C. Bishop	2	2	..	1	..	..	..	..	..	16	19
	Henry Steers	12	2	..	5	..	..	..	1	..	16	36
	H. Wallace & Percy Pyne	7	1	..	3	..	1	..	3	..	4	19
17	Henry Steers	..	..	..	..	..	..	..	..	..	..	..
	Francis C. Bishop	2	2	..	9	..	3	..	10	..	7	33
	H. C. Wallace-Percy Pyne	6	1	..	15	..	..	..	..	..	23	45
19	H. C. Wallace-Percy Pyne	1	..	..	1	1	..	..	..	1	2	5
	Henry Steers	..	..	..	..	..	..	..	2	..	1	4
20	Henry Steers	..	..	..	..	..	..	..	..	1	..	1
1919 Jan. 2	Spencer Aldrich	8	..	1	2	3	..	..	3	..	4	21
	Lt. S. W. Aldrich	2	..	..	18	4	..	..	8	..	3	35
3	Lt. S. W. Aldrich	4	2	..	18	4	..	..	9	..	..	37
	Spencer Aldrich	3	3	1	4	1	..	..	1	..	..	13
13	E. A. Carter	3	5	2	3	..	..	..	2	..	..	15

Information from one page, game log 1918-1919 season. First season entry November 4, 1918; last on January 28, 1919. Season total: Man-days hunting, 70; total bag, 1048; average daily bag, 14.9.

GAME REGISTER—Pochantas Fowling Club, Inc.—(Continued)

1934 Date	Name	Black	Mallard	Sprig	Widgeon	Ruddy	Black head	Red head	Canvas	Misc.	Geese	Total
Dec. 1	John Paine	2	1	3	..	1	..	..	..	..	..	7
6	E. A. Carter	..	1	11	..	..	..	..	..	..	..	12
	Spencer Aldrich & son	1	..	5	..	..	..	..	..	..	..	6
7	E. A. Carter	2	..	10	..	..	..	..	..	..	1	13
	Spencer Aldrich & son	..	..	3	..	..	..	..	..	..	4	7
8	E. A. Carter	1	1	10	..	..	..	..	..	..	..	12
	Spencer Aldrich	1	3	1	..	..	..	..	..	..	..	5
13	E. M. Allen	..	..	..	..	..	..	..	..	..	..	..
14	E. M. Allen	4	1	1	..	..	..	..	..	..	..	6
15	E. M. Allen	1	..	1	..	..	..	..	..	..	..	2
20	R. Cheyney	1	..	11	..	..	..	..	..	..	..	12
21	R. Cheyney	1	..	3	..	..	..	..	..	..	..	4
22	R. Cheyney	..	1	2	2	..	..	..	..	..	..	5
1935 Jan. 3	E. A. Carter	..	..	4	2	..	..	..	..	..	..	6
	C. Goodwin Carter	..	..	1	..	..	..	..	..	..	..	1
4	C. Goodwin Carter	..	..	5	..	..	..	..	..	..	..	5
	E. A. Carter	..	..	12	..	..	..	..	..	..	..	12
5	E. A. Carter	1	..	6	..	..	..	..	..	..	..	7
	C. Goodwin Carter	4	..	5	..	..	..	..	..	..	..	9
11	Spencer Aldrich & son	6	..	5	..	..	..	..	..	..	..	11
	E. M. Allen	2	..	6	..	1	..	..	2	..	..	11
12	E. M. Allen	3	..	1	..	..	..	..	..	..	..	4
	Spencer Aldrich & son	1	1	4	..	..	..	..	..	..	..	6

Information from one page, game log 1934-1935 season. First entry November 8, 1934; last on Jan. 12, 1935. Season total: Man-days hunting, 37; total bag, 307; average daily bag, 8.

1948 Date	Name	Black	Mallard	Sprig	Widgeon	Ruddy	Black head	Red head	Canvas	Misc.	Geese	Total
Dec. 10	Frank H. Redwood											
	A. B. Hodges	1		6				1			2	10
	K. K. Wallace									7		12
	W. G. Beaman				1		4					8
	C. W. Eley						7	1				8
	A. P. Jones											
	L. D. Williams											
11	Frank H. Redwood											
	W. G. Beaman	1		3			1			3		8
	A. B. Hodges	1		1	2		1			3		8
	C. W. Eley											
13	Curtis Turner						3			2		5
	Bob Allen	1		1								2
	A. W. Beaman											
11	Leigh Williams											
	Arthur Jones	1		3			2			6		12
	K. K. Wallace											
13	K. K. Wallace											
	Dr. Walter Adams				1			1		1		3
14	H. D. Clarke											
	A. B. Hodges				2		1	4			1	8
15	Tommy Thompson		1		3		1			2		7
	Arthur Jones											
15	Wm. G. Redwood											
	Frank H. Redwood			1	7						1	9

Information from one page, game log 1948-1949 season. Season opened December 10, 1948; closed Jan. 8, 1949. Season totals: Man-days hunting, 97; bag, 266; average daily bag, 2.5.

CHAPTER II

The Aftermath

An entire volume of such quotations could readily be compiled. The West coast, the Atlantic tidewater, the marshlands bordering the Gulf of Mexico, the potholes and sloughs throughout the length and breadth of the Mississippi and Missouri River drainages—all have their background of waterfowl abundance. Also, they all have an unsavory record of ruthless butchery and wanton waste by sportsmen and market gunners alike. These sketches, however, serve no useful purpose other than to point up the fact that the opportunities for duck and goose shooting in years gone by—years that will never return—were in startling contrast to what we find today.

When the marshes and lakes were blackened with waterfowl and the forests and plains abounded with deer, elk, and buffalo, the early American settler had bred in him a persistent idea and a wasteful philosophy. He became imbued with a firm belief that the wildlife resources were inexhaustible, that extermination was impossible. That philosophy resulted in wasteful disregard for all forms of wild creatures. "After all," he thought, "these things will always be here, so why should anyone try to curb my inborn love of hunting?" That idea and that philosophy became ingrained to the point that it exists even in the present generation. Also, there was handed down a firm conviction that the native American has a birthright of free hunting to be exercised pretty much where and when he pleases.

But what has been happening to the environment that produced and fed and protected those millions of waterfowl in the early days of our history?

Now, in the great Central Valley of California—the marshes

Plate 4. "Three Canvasbacks" over a typical marsh was the scene depicted in the second duck stamp, by Frank W. Benson, in 1935. Sale 448,204.

that Andy Burnett rode through in the fall of 1832—there are rice fields, orchards, and vegetable farms. Now, California has few places where ducks and geese are not looked upon as pests by the agricultural interests. Those few spots consist of pitifully inadequate state and Federal refuges, and scattered privately-owned gun clubs. More and more lands are being converted to agricultural uses and habitat is steadily declining. Yet, the opposition of a few influential but short-sighted sportsmen in California has made the expansion of the refuge program there most difficult. They resist any move to set aside more areas for the birds because they feel that this will reduce their own opportunities to hunt wherever they please.

The story is much the same in other parts of the country. Throughout the vast reaches of the grassland country extending from

Montana to Minnesota and southward through Nebraska and Iowa, the mighty hand of agriculture has squeezed the water from the marshes and the potholes until there is now only a semblance of the hundreds of thousands of sloughs that were there only 50 years ago.

Thirty-eight states, all in fact, except Connecticut, Maine, Massachusetts, New Hampshire, New Jersey, New York, Pennsylvania, Rhode Island, Vermont, and West Virginia, make legal provision for drainage enterprises. The result has been that some ninety million acres have been tiled and ditched since 1900. Nearly two-thirds of the lands in drainage enterprises are located north of the Ohio and Missouri Rivers—the area that was once the principal producer of waterfowl in the United States. Much of the other one-third lies along the lower Mississippi and in the regions bordering the coast lines of the southern Atlantic and the Gulf of Mexico. Here we find large-scale drainage undertakings in the vast marshes and swamps that were once the principal wintering grounds for the birds east of the Rocky Mountains.

Rainfall that once filled potholes, sloughs, and ponds, and overflowed into quiet, shady brooks now is hurried through drain tiles to straight ditches and into streams that are being straightened and deepened as rapidly as funds become available for the control of floods. Now those waters rush with ever-increasing speed into the Ohio and the Missouri, and finally the Mississippi Rivers. Little wonder that the floods in the lower Mississippi River seem to be unmanageable. Yet, a law passed by the 80th Congress authorizes the study of another sixty-five million acres of wet lands looking toward drainage to reclaim them for agriculture.

We Americans, in our zeal to improve our physical well-being so as to have better homes and offices and roads and bank accounts, are inclined to overlook the heritage which Nature gave to us more than to any other nation in the world. Whenever there is found a fine stand of timber, there is also certain to be someone nearby who wants to cut it—usually with little thought of the needs of the future. If there is a marsh filled with cattails or wild rice, muskrat houses dotting its surface, bullfrogs croaking and blackbirds singing, ducks and geese paddling about with broods of downy young in their wake, certainly there is also someone busy with pad and pencil figuring how much it would cost to run a drainage ditch through the heart of that marsh. Someone is estimating how many farms could be sold if that lake were eliminated—and how much profit might be made from the transaction.

On the other hand, if there is a clear, shade-covered stream tumbling out of the hills, cool and inviting to smallmouth, to trout, to wood ducks, someone else is certain to be busy with another pad and another pencil figuring how much it would cost and how much profit could be made from power or from irrigation if a dam were thrown across the narrows and the stream converted into a lake. We are never satisfied. We are always trying to improve upon Nature. Oftentimes wildlife conservationists would much prefer to leave Nature alone. They are always fighting what often seems to be a losing battle.

Canada has experienced the same change. Agricultural expansion in the prairie provinces of Canada—Alberta, Manitoba, and Saskatchewan—has proceeded even more rapidly than in our own prairie states. South Central Canada was formerly a veritable cradle of waterfowl production, but the plow there has made the same serious inroads on nesting habitat that it has in Nebraska, Iowa, the Dakotas, and in western Minnesota.

At the turn of the century there were approximately 3,600,000 acres under cultivation in Alberta, Saskatchewan, and Manitoba; by 1920 this area had increased to 31,758,000 acres; by 1930 to 33,156,000 acres; and by 1947 to nearly 43,500,000 acres. This means that 68,000 square miles, an area roughly the size of the state of Washington, has been converted from grass-bordered prairie potholes and sloughs—duck habitat—to grain and summer-fallowed fields. This evolution has taken place in the very heart of the principal duck nesting area in Canada. Of course, farming has not driven all of the ducks and geese out of the country, but it has placed consistent production in real jeopardy. Agriculture has been responsible for the drying up of many bodies of water, while spring burning of stubble fields has destroyed many nests, as also has early plowing. John Lynch, one of the Fish and Wildlife Service biologists working in Saskatchewan, one spring counted 23 duck nests that were certain to be p'owed under in a half-section of land. Undoubtedly he missed many others.

The impact of the expansion of agriculture and industry on waterfowl goes much deeper than the mere drainage of marshes.

Some drained marshes—and there are many of them where the sour peat soils were unfit for farming after the water had been removed—can be restored. But, warranting real concern has been the lowering of the ground water levels which has occurred in many areas of the United States. With underground water reservoirs being sucked dry to provide the things we demand for our present standards of living, surface depressions lose their water levels much more quickly than they did formerly. In those areas, instead of spring rains and thaws filling potholes that will remain stable through the hatching season, there is an increasing tendency for those same waters to disappear so that marshes now go dry in early summer. Then we have appalling losses of fledgling young and moulting adults. The factor of safety is now so narrow during the breeding seasons that drought conditions on the vast prairie breeding areas of Canada and our own United States immediately result in drastically curtailed production. The direct return to the hunter is inevitable—a small bag limit and a short season.

This lowering of the ground water level is much more serious than most of us realize. It comes about because our modern civilization demands more from the underground rivers than Nature returns to them from the skies. Did you ever stop to think about how much water we use—and waste—in our daily lives? The average American requires more than 100 gallons of water each day, most of which is needed by industries or farms that support his community. Europeans use only a quarter as much. Do you realize that it takes 65,000 gallons of water to produce a ton of steel, seven to ten gallons to produce one gallon of gasoline, and fifteen gallons to make one gallon of beer?

Lester Velie, discussing some of the critical areas in a thought-provoking article published in *Collier's* (May 15, 1948), reports that:

At Long Beach, Calif., thirsty war factories and a mushrooming community gulped so much water from underground pools that water levels, once above the sea, are now 75 feet below sea level. As the fresh water retreats, the ocean seeps into the water-bearing beds, polluting wells.

The Middle West and the East are also drawing down their underground supplies. Water levels under the downtown district of Louisville, Ky., have dropped 40 feet in ten years. The Indianapolis water table is down 50 feet. Air-conditioning systems have been gulping the city's water and are posing this question for the rest of America, too: What to do about it?

Baltimore had to reduce pumping at the beginning of the war when wells became contaminated with salt. Philadelphia's water tastes more like Epsom salts every year. Overpumping has pulled Brooklyn's water down 35 feet below sea level, giving it a rank, musty taste.   *   *   *   *

Townsfolk in Santa Barbara, forced to ration water, rename their city Sahara Barbara. The well-to-do import water by truck to save their lawns, paying $40 per two-hour wetting. Vigilantes check meters for water hogging.

At Los Angeles, County Supervisor Raymond V. Darby asks the California legislature to put up $1,000,000 for ideas on how to get sweet water from the sea, cheap.

In California's Central Valley, farmers slaughter their dairy cows and send steers out of the state by the thousand. No pasture. Beekeepers ship out their bees. No flowers.

The kind of water shortage many communities face has nothing to do with drought. The fact is we are using up our underground water faster than nature can replenish it. (More than half of the water we use daily comes from underground.)

In the Southwest, experts are already reckoning the day when the last newcomer will cross the Rockies heading west, because there won't be enough water for all. Some Los Angeles engineers have even set the deadline—1968.

All of these factors—agriculture, industry, the pressure of increased human populations, has not been good for waterfowl. If you doubt that statement, why not draw on your own experience? Do you remember the marsh where your Dad taught you how to hold on that first mallard you ever shot? How about visiting that same spot the next time you get a chance? The odds are about 30 to 1 that you will find on that spot, instead of a marsh filled with birds, a field of corn, wheat, rice, or cabbage. Or you may even find an apartment building or a filling station on it; or perhaps a highway or an airport running across it. That spot will never produce, nor feed, nor rest another duck. That spot symbolizes American economic progress. That spot is a reminder that the waterfowl program is in competition with a host of other uses of land and water, and that if the resource is worth preserving, sportsmen and conservationists *must* be willing to pay prices sufficient to compete with those other demands. They *must* insure that suitable areas are acquired, developed and maintained for the use of waterfowl. As "Ding" Darling said years ago,

"Ducks don't mate on the wing, and they don't build their nests on the tops of fence posts." There is no other answer than the establishment of enough suitable areas to provide permanently for at least the minimum needs of a reasonable continental waterfowl population.

Many drainage projects have been failures, and wherever possible they should be restored to wildlife. Many other areas that are now outstanding for wildlife should be perpetuated in their present state. To preserve them, wildlife interests—Federal, state, and private—must acquire them and keep them; otherwise, they will go the way the California lowlands have gone—the way the Horicon, the Kankakee, and the Mingo Marshes lost their life-giving waters.

It does little good for conservationists to condemn and decry and blame other folks. If the great sport of wildfowling becomes a thing of the past, they will have only themselves to blame. The sad part of it is that it will be their youngsters who will suffer the greatest loss. It is they who will point to the short-sightedness and selfishness of their sportsmen dads who let such things happen. Unless conservationists bestir themselves, unless they begin to think seriously about preserving and restoring habitat, the fight will surely be lost. Regulations alone can prolong the sport of duck hunting for a while, but not for long. Sportsmen must worry more about the needs of the birds and less about their privileges—longer seasons, bigger bag limits, the return of bait and live decoys, and other improved methods of killing. Otherwise, there soon will be no need to worry about the annual regulations because there will be nothing left to regulate.

CHAPTER III

The Slow Awakening

As EARLY AS 1776, the colonies of New York and Massachusetts showed interest in wildlife to the point that their governing bodies enacted a few regulations for the protection of wildfowl, but elswhere for many years thereafter little thought was given to game laws. In 1848, Massachusetts passed a law governing the netting of passenger pigeons, but it was not to protect the birds; rather, it extended the long arm of the law to punish anyone caught frightening the birds out of the huge nets set by the market hunters. In 1857, Ohio considered legislation to give these fine birds some protection, but it failed of passage because of the cogent argument that "no ordinary destruction can lessen them or can they be missed from the myriads that are yearly produced."

In spite of feeble attempts to recognize a changing situation—changing constantly for the worse—wildlife destruction continued apace. The last wild elk in New York was killed in 1845. The wild turkey disappeared from a large part of that state at about the same time, and some 15 years later the last moose in the Adirondacks was gone. Spring shooting of waterfowl was forbidden in Rhode Island in 1846, but the law was later repealed. In the 1830's, New York and Virginia attempted to outlaw the swivel punt guns and other devices that were designed for the mass killing of rafted waterfowl.

The first prohibition against wanton destruction and waste of game appeared in the West in Washington Territory in 1865, and in Wyoming and Colorado in the early 70's. In 1864, a closed season on buffalo hunting was enacted in Idaho—the first closed season on an animal that had been thought to be countless and inexhaustible. Between 1850 and 1885 game legislation began to receive its first real consideration generally, and by 1880 there was some sort of legal protection for wildlife in all of the states and territories. The bulk of these were significant in so far as waterfowl were concerned because even then it was recognized that spring shooting and other destructive methods of killing needed to be curbed.

The 19th Century saw a great emergence of students of nature. That period produced the works of such noted artists and naturalists as Audubon (1780-1851), Wilson (1766-1813), and Baird (1823-1887). It witnessed the joining together in small groups and associations of earnest observers, the forerunners of our present modern scientific societies for the study, protection, and preservation of American wildlife. The Nuttall Ornithological Club, formed in Cambridge, Massachusetts, in 1873, was the first organized vehicle for the companionship of numerous individual nature students and for the exchange of notes.

Actually, the American Ornithologists' Union, an offshoot of the Nuttall Ornithological Club, might well be said to be the beginning of the U. S. Biological Survey, since its work led to the formation of that Government unit. The work of three of the Union committees, those on the English sparrow, on bird migration, and on the study of faunal areas, brought such widespread interest that, being unable to handle the volume of work, the Union addressed a memorial to the Congress with the result that the Agricultural Appropriation Act of 1885 included $5,000 of Federal funds for work on economic ornithology under the then Division of Entomology, which later became the Bureau of Entomology and Plant Quarantine. Dr. C. Hart Merriam, Chairman of the American Ornithologists' Union's Committee on Migration and Geographical Distribution of Birds, was selected as the first chief of the new unit set up to study birds. So began the Government unit which was later officially designated as the Bureau of Biological Survey and which was expanded to undertake many new functions in the field of wildlife research and conservation as the march of progress brought ever-changing conditions and ever-new problems. On June 30, 1940, this bureau was combined with the Bureau of Fisheries, formerly of the Department of Commerce (originally the United States Fish Commission organized in February 1871), and became the present Fish and Wildlife Service of the United States Department of the Interior. During all that time the Government has increased its efforts to learn about the habits of birds and to give added protection to those species that need it.

Wild Things in Relation to Agriculture

Agricultural interests actually played a more important role in early legislation dealing with wildlife problems than did those now represented by conservation organizations. In the early 70's, outbreaks of Rocky Mountain locusts in the West and Southwest brought great destruction to the crops of farms and ranches. It was noted at that time that birds played the most important role in keeping these ravenous little insects from wreaking complete devastation.

As early as 1896 the California State Board of Horticulture had strongly recommended a national law which would regulate bird and animal importations. It was the first state to take a determined stand in favor of severe restrictions on foreign importations of this nature. And for good cause. California's budding horticultural and vegetable industries were extremely vulnerable to the introduction of animal species from the Orient and South Pacific that might prove to be plant pests. Since 1883 the state had prohibited the landing, and authorized the confiscation at California ports, of flying foxes, Australian wild rabbits, mongooses, and other creatures of dangerous possibilities. The United States can probably attribute the present freedom from these pests to this wise and timely action by the State of California. Hawaii, Puerto Rico, and the Virgin Islands—all American territories—are over-run with the mongoose which was introduced in the mistaken belief that this vicious little brute would exterminate rats. Destructive as is the rat, the mongoose is worse.

Other sections of the country also were beginning to worry about the unwise introduction of exotics. An English sparrow resembles a mallard duck about as closely as a prairie dog resembles a moose, yet this lowly little bird had much to do with the early history of legislation which later became the basis for managing waterfowl. Back in the 50's several enthusiastic sponsors of a movement to supplement our native fauna succeeded in importing the European house sparrow, thinking that these little birds would aid agriculture by suppressing the geometrid caterpillar. The sparrow liked our shores, as amply attested by his immediate response and spread. He does eat the geometrid caterpillar to some extent but, unfortunately, his aggressiveness over other bird life and his nuisance value soon far outweighed his virtues. He refused to stay in New York, where originally introduced, but rather multiplied and overran many other areas. This soon pointed up the futility of any one state attempting to regulate a species which knows no political boundaries.

The last half of the 19th Century seems to have been a period when there was great interest in attempting to improve on Mother Nature by introducing various foreign species. This naturally led to wide differences of opinion between those who favored the native fauna and those who advocated bringing in exotics from other lands. The American Society of Bird Restorers was organized in Boston at the end of the last century to oppose the growing trend toward the stocking of America with foreign birds. It was composed of members who were convinced that native birds should be given a chance to resume the place in the sun which they had originally held. Boston was a logical place for such a group to come together because that city always has been a center of interest in birds. Also, its Commons and Public Gardens were no longer crowded with the lovely and songful native species. Instead, the ivy-covered buildings were now the homes of a raucous horde of chattering and chippering sparrows, which fouled the landscape and added insult to injury by dropping their filthy, whitish pellets on the sidewalks and passersby beneath. Members of the Society, with outraged dignity, decided to take action against those foreign intruders and determined to rid the city of them.

This society in all good faith immediately set about to undo the mistakes of those people who, with equal good intent, had introduced the troublesome English sparrow. They employed crews who with ropes and ladders immediately proceeded to destroy the nests and the young birds. They cleaned the cornices of filthy nests, they plugged every hole that would harbor a sparrow, they killed the young, and they broke the eggs. But not for long! Soon the humanitarians and the others who could not bear the thought of killing any form of bird life arose in righteous anger and descended on the mayor of Boston with the result that that gentleman was obliged to rescind the permit that had been issued to the Society. Thereupon, their task of ridding the city of this pestiferous nuisance came to an abrupt halt.

This incident, with its attendant publicity, was soon a matter of discussion wherever newspapers were read. It focused the attention

Plate 5. Richard E. (Dick) Bishop, won the award in 1936 with this drawing of "Three Canada Geese" on the wing. Sale 603,623.

of the public not only on the episode of the Boston Commons, but also on the unknown dangers that always attend the introduction of a foreign species into a new land. The first bulletin of the Biological Survey, entitled "The English Sparrow" soon came into great demand throughout the entire country. Published in rather optimistic numbers in 1888, it had soon, as is the case with many Government publications, run its course of popularity, and as a result, there were still large numbers on the shelves and the stock was not moving. This situation was changed overnight by the Boston Commons affair. The leftovers of this famous old bulletin were sent out in large numbers, and within two weeks the supply was exhausted with a long list of unfilled requests.

People all over the country began to worry about how soon the English sparrow pest might invade their cities and crowd out the native birds they loved so well.

The English sparrow incident happened at a time when the public was developing a frame of mind which recognized the doubtful wisdom of haphazard importation. Along with it there was a growing consciousness of the need for better protection of native species. People were beginning to worry about the unrestricted killing that had brought many native forms to the verge of scarcity. The time was ripe for some concerted action designed to remedy the situations that were perplexing the agriculturalists, the conservationists, and the sportsmen of that period.

CHAPTER IV

Protective Legislation

*B*ETWEEN 1897 and 1900 three different phases of the wildlife conservation movement were given Congressional consideration. With what might now be considered as prophetic vision, inasmuch as the Government's interest in fish and wildlife has been consolidated into the present Fish and Wildlife Service, a western Congressman, long noted as a game bird enthusiast, proposed to give the United States Fish Commission jurisdiction over game birds. He wished to authorize the Government to engage in such activities as restocking barren covers, establishing game birds that would be suited to certain sections of the country, and importing birds from foreign lands. Another bill introduced by a western Senator within a few days of this one proposed to prevent the illegal export of big game from Colorado, Utah, and Wyoming. Later, in the same Congress, a Senator from Massachusetts introduced a bill which would control the traffic in bird plumes, both importations and those taken within any state.

The Lacey Act

None of these bills became law but in the following Congress a new bill was introduced which included all of the features of the three that had been introduced in the previous session of the Congress. In summary, the bill proposed to suppress the killing of game as a business, a form of destruction then popularly known as market hunting. It also proposed to make much more difficult the taking of plumes and feathers from both game and non-game species; to regulate the introduction into this country of all exotic species of birds and mammals; and to prohibit by law the introduction of species which were known to be injurious to American wildlife or to agriculture. In short, this new bill was designed to improve and enhance the status of wildlife by providing added protection and by excluding foreign competitors. The interesting thing is that this bill, drastic as it was for times such as those, met with little opposition. It was passed on May 25, 1900, and has since become universally known as the Lacey Act. Enforcement of this Act was not placed with the Fish Commission, as was originally proposed, but with the Department of Agriculture, and as such became a part of the duties of the Biological Survey.

From the standpoint of over-all wildlife conservation, the Lacey Act was one of the most important measures ever to pass the Congress. It has since become one of the foundation stones for good wildlife management and continues to be used actively and regularly to enforce the wildlife protective laws of the states and of the Federal Government. The Lacey Act, as amended in recent years, is not a mere "pat on the wrist" act. It has teeth. Legal authority stems from the powerful interstate commerce clause of the Constitution. Among its provisions is a prohibition against the shipment of game taken illegally in one state and transported across state boundaries contrary to the laws of the state where taken. This means that illegal shipment or transportation of game, or parts of game, or of other protected species from one state to another becomes a Federal offense. Anyone doubting the effectiveness of this law should examine the court records, particularly those pertaining to the illegal shipment of beaver and other furs from one state to another.

When the Act was first passed there were not many state laws that governed the shipment and handling of game, but this situation has constantly changed so that now the Lacey Act is one of the most potent tools for the enforcement of state wildlife laws. To quote an early court decision, the object of the Lacey Act is to "enable the states by their local laws to exercise a power over the preservation of game and song birds, which without that legislation, they could not exert." This act is not designed in itself to strike at the illegal shipment of game or the traffic in bird plumage, but rather it brings to bear that very potent interstate commerce clause which enables the states to do the striking for themselves.

It was high time that such a law should have been enacted. For 15 years prior to the passage of the Lacey Act, the taking of wildlife for commercial purposes had been a national scandal. Forty thousand terns are said to have been killed around Cape Cod, Massachusetts, in the very year the American Ornithologists' Union was organized. These beautiful little birds had already been practically extirpated on the New Jersey coast. Herons, ibises, egrets, gulls, roseate spoonbills, and other non-game birds of fine plumage were suffering excessive decimation all along the Atlantic and Gulf coasts. Even in California, Oregon, and Washington, plume hunters were plying their nefarious trade. The rookeries in Florida suffered the greatest destruction, as regular squads of paid hunters were maintained by dealers and local contractors. Purchasers regularly shipped enormous quantities of bird plumes to New York and other centers of traffic. And this, despite the fact that since 1877 Florida had had a law protecting plume birds.

Out of this traffic in wildlife came the inspiration in 1886 for the organization of the Audubon Society formed "for the protection of wild birds and their eggs." The Boone and Crocket Club was formed a year later. The League of American Sportsmen was organized in 1898 with the avowed purpose of urging more adequate enforcement of game laws and better protection of insectivorous and song birds. It played an important part in the passage of the Lacey Act. All of these organizations developed into militant groups that were pledged to the cause of wildlife conservation.

No national legislation—particularly dealing with wildlife management—is ever spontaneous. The introduction of a bill into Congress may have been sparked by a sudden impulse, but the end product has been well refined by the heat of discussion and controversy before it ever reaches the Chief Executive for final action. This was particularly true of the Migratory Bird Treaty Act. That bill represented Conservation's first great effort to pull its boots from the mire of complacency and to shake off opposition from the ranks of the market hunters and the greedy plume gatherers. That emergency was due to a mighty force from within—a force that had been building for the day of awakening to the fact that the supply of waterfowl was not inexhaustible. The Lacey Act, although representing a long step forward, was proving to be insufficient. Spring shooting, market hunting, and other organized forms of destruction proceeded apace.

Greedy game markets in Boston, New York, Philadelphia, Washington, Baltimore, Chicago, New Orleans, Salt Lake City, San Francisco, Portland, and Seattle readily accepted game of all sorts as shipments came from the great slaughtering grounds. No section of the country that had a good supply of wildlife escaped. Expert hunters, working six and sometimes seven days a week, from daybreak to dark, killed every game bird within reach. On one club in southern California in 1906 two hunters, with automatic shot guns, are reported to have killed 218 geese in an hour, with a total for the day amounting to around 450. In recording his experiences, one of these men remarked that because of the warm weather they had considerable difficulty in keeping the birds from spoiling!

From other areas there are records of heavy kills by the market hunters of the early days. For instance, one such hunter on Marsh Island in Louisiana killed 430 ducks in a single day. The slaughter inflicted by a 4-gauge shotgun as big as a small cannon was terrific. With "roman candles" loaded with a half-pint of black powder and a hat-full of shot, the hardy old market hunters could literally keep the air full of duck feathers.

One morning before 9 o'clock, one of the champion duck killers along the Mississippi River shot down 122 wood ducks, the handsomest of all ducks that are now so scarce that they are given complete protection in many states. He also claimed to have cut down 120 blue-winged teal in just that many minutes. During one camping trip he shot, barreled, and delivered to market a total of 1,920 assorted waterfowl. Someone told him about the swarming bird life over the North Dakota prairies, so he made a special trip out there with his "Big Bertha." On the first day, between 3 o'clock in the afternoon and sundown, he brought down 46 Canada Geese and 37 Sand Hill cranes. The 700-odd pounds of birds were a heaping load in a two-horse buckboard. Shooting was so profitable that he put in a solid month of it before heading back east again.

Another noted market hunter of the Mississippi delta claims

that during his years of shooting he probably killed 100,000 ducks and geese which he sold at prices ranging from 7 to 10 cents each. His shooting was ruined when the Government finally imposed a daily bag limit of 25 birds.

In Currituck County, North Carolina, during the period from 1903 to 1909, according to T. Gilbert Pearson, long the Executive Director of the National Audubon Society, local buyers paid not less than $100,000 annually to the 400 duck hunters of that region, the fowl generally being shipped to Baltimore, Philadelphia, and New York. Prices varied, but at times gunners received $1.60 a pair for redheads, and $2.75 a pair for canvasbacks. Canada geese were worth 40 to 50 cents each and ruddy ducks commanded 90 cents a pair. A Currituck Sound duck hunter and three companions in the month of November, 1905, received $1,700 for 2,300 ruddy ducks that they had shot and sold for shipment. One hunter reported that he personally shot 282 ruddy ducks in one day. On another occasion he killed 75 canvasbacks, 5 swans, 4 Canada geese, "and a few other ducks." His ammunition gave out and he stopped shooting before the day was over.

A duck buyer at Norfolk, Virginia, had a storage warehouse and throughout the shooting season would sail his boat to the different gunners who were operating in sink-boxes, batteries, or blinds and buy their ducks. These he brought to Mill Landing, packed them in barrels of ice, and sent them to Norfolk by wagons, often as many as 1,000 birds at a time. In the early years he paid 40 cents a pair for most species—four ruddy ducks were always counted as a pair. Redheads and canvasbacks brought from $1.00 to $1.50 a pair. Geese he bought for 50 cents apiece. He seemed to have had pretty much of a monopoly on the duck-shipping business in Back Bay from 1888 until the sale of ducks was prohibited by Federal law on July 3, 1918. His harvest of ducks began early in October and continued for six months every season for thirty years.

The newly-enacted Lacey Act had not stopped the illegal shipment of game from one state to another. There was much profit in the market hunting business and few officers to enforce the law. Ruffed grouse and quail were found packed in butter firkins, with a solid layer of butter on top, labeled, of course, as "butter." Others found their way into the markets marked "eggs." Game was carried or checked on trains in suitcases and trunks. Some wardens, when a traveler refused to open his luggage for inspection, joyously tested the contents by boring an inch hole through the bottom.

William T. Hornaday tells us:

> Once upon a time, a New York man gave notice that on a certain date he would be in a certain town in St Lawrence County, New York, with a palace horse-car, "to buy horses." Car and man appeared there as advertised. Very astentatiously, he bought one horse, and had it taken aboard the car before the gaze of the admiring populace. At night, when the A. P. had gone to bed, many men appeared, and into the horseless end of the car they loaded thousands of ruffed grouse. The game warden who described the insident to me said: "That man pulled out for New York with one horse and half a carload of ruffed grouse!"

The Weeks-McLean Law

But the day of such excessive killing was bound to come to an end. Public sentiment was becoming more and more aroused to the need for a more effective means of protecting wildlife, particularly migratory waterfowl. In December, 1904, a bill known as the Shiras Bill, designed to protect the migratory birds of the United States by giving additional authority to the Federal Government, was introduced in Congress. It failed of passage but for the next nine years there was increased agitation by the friends of wildlife so that Congress was not permitted to forget the dire situation into which waterfowl were being forced by the uncontrolled killers.

The Weeks-McLean Law, which contained many of the essential features of the Shiras Bill, became effective on March 4, 1913. This Act stated boldly, "All wild geese, wild swans, brant, wild ducks, snipe, plover, woodcock, rail, wild pigeons, and all other migratory game and insectivorous birds which in their northern and southern migrations pass through or do not remain permanently the entire year within the borders of any State or Territory, shall hereafter be deemed to be within the custody and protection of the Government of the United States, and shall not be destroyed or taken contrary to regulations hereinafter provided therefor." The Department of Agriculture was given the authority to make and enforce regulations which would give adequate protection to the enumerated wildfowl, and the law provided that any person who violated any regulation prescribed by the Secretary should be found guilty of a misdemeanor and fined not more than $100 or imprisoned not more than 90 days, or both, in the discretion of the Court.

To many, such a radical invasion of the inherent rights of American citizens to kill when and where and as much as they pleased was utterly preposterous! Imagine the Federal Government attempting to take away those privileges that had been handed down by fathers, grandfathers, and great grandfathers! Outrageous! Soon the opposition began to organize, but at the same time the friends of wildlife rallied to support the new law. The Weeks-McLean Act actually was on very weak ground from a legal standpoint. It was attached to an appropriation bill for the Department of Agriculture, and legislation in an appropriation act has always been on thin ice. Furthermore, there was grave doubt that a bill of this sort was constitutional.

The backers of the Weeks-McLean Bill, knowing the law had a constitutional weakness and realizing that its foes would never rest until they had taken the disputed measure to the highest court of the land, wisely took another course. A resolution was adopted by the same session of the Congress which requested the President to initiate negotiations with other countries of the North American continent looking toward the establishment of international treaties. A treaty draft, drawn with the aid and assistance of the leaders of the American Game Protective Association, was submitted to officials of the United States and of Great Britain. World War I interfered for a time, but eventually the treaty was concluded by representatives of Great Britain and the United States in Washington on August 16, 1916. The President of the United States formally approved it on September 1, and Great Britain ratified it on October 20. After exchange of the ratifications, the treaty was promulgated by Presidential proclamation on December 8, 1916. Eighteen months later, on July 3, 1918, the Migratory Bird Treaty Act became law. As was expected, the Weeks-McLean Law had been attacked on grounds of constitutionality and reached the Supreme Court of the United States. The case was dismissed in January, 1919, on motion of the Attorney General, since the Migratory Bird Treaty Act had made a decision unnecessary.

The Migratory Bird Treaty Act

Where the Weeks-McLean Law stated that migratory game and insectivorous birds were henceforth to be considered within the custody and protection of the United States Government, and should not be taken contrary to regulations of the Department of Agriculture, the new Migratory Bird Treaty Act went considerably beyond these provisions in its sweeping scope. It stated that *"unless* and *except* as *permitted* (italics supplied) by regulations made as hereinafter provided, it shall be unlawful to hunt, take, capture, kill, attempt to take, capture or kill, possess, offer for sale, sell, offer to purchase, purchase, deliver for shipment, ship, cause to be shipped, deliver for transportation, transport, cause to be transported, carry or cause to be carried by any means whatever, receive for shipment, transportation or carriage, or export, at any time or in any manner, any migratory bird, included in the terms of the convention between the United States and Great Britain for the protection of migratory birds concluded August sixteenth, nineteen hundred and sixteen, or any part, nest, or egg of any such bird."

It implemented the treaty which specified that there should be a closed season on migratory game birds between March 10 and September 1, with certain limited exceptions. The treaty provided that the open seasons prescribed should not exceed three and one-half months, that there should be complete protection of migratory insectivorous birds, and in addition that for a ten-year period there should be no hunting of certain rare birds, such as bandtailed pigeons, little brown, sandhill and whooping cranes, swans, curlews, and with certain minor exceptions, all shore birds. It provided special protection for wood ducks and eider ducks. The Migratory Bird Treaty Act made illegal the sale of migratory birds, the practice that had been even more destructive than spring shooting.

Those who drafted this legislation were farsighted. The law not only established restraint over the hunters by authorizing limits of kill and means by which game might legally be taken, but it also provided for controls over the birds. The Act permits the destruction of birds that may be damaging agricultural crops or other property, all under permits issued by agents of the Government. It authorizes the taking of birds for scientific and propagating purposes. Finally and all-important, it authorizes the making of regulations to meet changing conditions so that the waterfowl and other migratory game birds may be protected in accordance with such situations as arise from year to year. While the Act does not spell it out in so many words, it was the clear intent of Congress that from July 3, 1918, forward and henceforth, the needs of migratory birds should be considered first and the desires of those who want to take them second. That has been the administrative policy since that date and

Plate 6. "Five Scaups" speeding over a windswept marsh were shown on the 1937 migratory waterfowl hunting stamp, drawn by J. D. Knap, of New York City. Sale 783,039.

it must continue to be so. Woe be to any Director of the Fish and Wildlife Service who permits the desires of the hunters to be placed above the welfare of the waterfowl.

The Act is As Solid as a Rock

As was to be expected, those who were being curtailed by the new law lost little time in initiating court action to test its constitutionality. The first case was a friendly suit. The constituted authorities of the State of Missouri instituted an action against Ray P. Holland who was then a law enforcement officer for the Biological Survey. Mr. Holland is well known in conservation circles for his many years as editor of *Field and Stream*. The Act was sustained on April 21, 1920, by a decision of the Supreme Court listed as the case of Missouri v. Holland, 252 U. S. 416. Portions of this decision are of such general interest at the present time that they are quoted herewith.

> The State, as we have intimated, founds its claim of exclusive authority upon an assertion of title to migratory birds, an assertion that is embodied in statute. No doubt it is true that as between a State and its inhabitants the State may regulate the killing and sale of such birds, but it does not follow that its authority is exclusive of paramount powers. To put the claim of the State upon title is to lean upon a slender reed. Wild birds are not in the possession of anyone, and possession is the beginning of ownership. The whole foundation of the State's rights is the presence within their jurisdiction of birds that yesterday had not arrived, tomorrow may be in another State, and in a week a thousand miles away. If we are to be accurate we can not put the case of the State upon higher ground than that the treaty deals with creatures that for the moment are within the State borders, that it must be carried out by officers of the United States within the same territory, and that but for the treaty the state would be free to regulate this subject itself. * * *
>
> Here a national interest of very nearly the first magnitude is involved. It can be protected only by national action in concert with that of another power. The subject matter is only transitorily within the State and has no permanent habitat therein. But for the treaty and the statute there soon might be no birds for any powers to deal with. We see nothing in the Constitution that compels the Government to sit by while a food supply is cut off and the protectors of our forests and our crops are destroyed. It is not sufficient to rely upon the States. The reliance is vain, and were it otherwise, the question is whether the United States is forbidden to act. We are of the opinion that the treaty and statute must be upheld. Cary v. South Dakota, 250 U. S. 118.

The validity of the Canadian "Migratory Bird Convention Act" was upheld by the Supreme Court of Prince Edward Island in a decision (Michalemas term, 1920) in the case of the King v. Russel C. Clark. Therefore, courts of both the United States and Canada have upheld the validity and power of the treaty and the subsequent enabling legislation of the two great nations. A similar treaty was negotiated with the United Mexican States, concluded on February 7, 1936, and ratified by the President on October 8, 1936.

Several cases have subsequently been brought to trial in which the authority of the Federal law has been questioned but in every instance it has been upheld. Some have challenged the authority of the Government to regulate the taking of waterfowl by prohibiting the use of bait. Others, the power to set seasons, while some have revolved about the right of the Secretary to close to the hunting of migratory waterfowl areas which included privately-owned lands. A brief review of some of the more important of these cases may serve to emphasize the sweeping powers granted to the United States Government under the treaty with Great Britain.

A motion by the Government in the Eastern District of Kentucky to dismiss an action by residents of Kentucky seeking to enjoin the U. S. Marshal and U. S. Attorney from enforcing the regulation prohibiting the hunting of mourning doves during all of September was sustained and injunctive relief denied on August 31, 1935, in the case of Shouse et al v. Moore et al, 11 Fed. Supp. 784.

A case involving regulations which prohibited the hunting of mourning doves and waterfowl by means of bait was concluded in favor of the Government in 1935 in the southern district of Georgia, U. S. v. Griffin, 12 Fed. Supp. 135. A similar successful case was decided in 1936 in the Southern District of Illinois, Brandenburg et al v. U. S. Attorney et al, 12 Fed. Supp. 342.

The Circuit Court of Appeals for the Seventh Circuit in 1937 affirmed the action of the Federal Court for the Southern District of Illinois which had imposed a fine of $200 and costs on two persons charged with taking migratory birds by means of feed or bait, Cochrane v. U. S. 92 Fed. (2d) 623. The United States Supreme Court denied the applications for writs of certiorari on January 31, 1938— 58 U. S. Sup. Ct.—Rep. 522.

A case from the Southern District of California (Cerritos Gun Club v. Hall, 21 Fed. Supp. 163) wherein it was contended the Migratory Bird Treaty did not authorize Congress to give the Secretary of Agriculture power to regulate the means by which migratory birds could be taken and in which the lower court decided in favor of the Government was appealed to the Circuit Court of Appeals for the Ninth Circuit (California). The appeals tribunal sustained the action of the lower court and in the course of its opinion, 96 Fed. (2d) 620, stated:

> We do not agree * * * that bringing the line of flight of wild fowl by baiting to a hunting territory is not indirectly a luring within the meaning of the regulation.
> We believe the appellants have violated the Secretary's regulation

whether by pursuing the indirect method of baiting before the season opens to keep the birds there to be shot after the season opens, so that hunters may flush them as they walk or punt over the preserves, or by directly placing the grain in front of the blinds or stands during the season. Wherever the grain is placed on the preserves, the wind will create lines of the birds' flight, to and from it, which will aid the slaughter from blinds located for the purpose.

Two cases relating to failure to charge the defendants with having knowledge of the presence of bait or feed in areas where mourning doves were being hunted were concluded in 1939. The court in a case in the Western District of Tennessee, U. S. v. Reese, 27 Fed. Supp. 833, decided:

The conclusion has been reached that, in the instant case, it is unnecessary for the Government either to aver in the information, or to prove at the trial, that the defendant had knowledge of the unlawful baiting of the hunting ground, in order to render him amenable to punishment for a violation of the statute and the regulations promulgated pursuant thereto by the Secretary of the Agriculture.

In the other cases in the Western District of Kentucky, U. S. v. Schultz et al, 28 Fed. Supp. 234, the court stated:

In view of the broad wording of the act, and the evident purpose behind the treaty and the act, this Court is of the opinion that it was not the intention of Congress to require any guilty knowledge or intent to complete the commission of the offense, and that accordingly scienter is not necessary. * * * *

Accordingly, the court is of the opinion that in each of the four cases under consideration the defendant is guilty under both counts of the information in question, even though there was no evidence of any guilty knowledge or intent upon his part at the time of the commission of the offense.

Early in November, 1947, residents of Illinois filed a petition in the Federal Court for the Eastern District of Illinois seeking a preliminary injunction against enforcement of a regulation closing certain areas in Alexander County, Illinois, to the hunting of wild geese in order to afford additional protection to the birds in this flyway. The request was denied.

The Migratory Bird Treaty with Great Britain and the subsequent enabling legislation known as the Migratory Bird Treaty Act are significant in that the destructive killing methods of earlier days were finally brought to a halt in a most forceful fashion. The hunting of waterfowl was no longer a right of the citizen. Now birds could be taken only as *permitted* by the regulations. If no regulations were issued, there would be no open seasons. Had it not been for the foresight and the persistence of those early conservationists there is little doubt that legal hunting of waterfowl as we know it today would have been a thing of the past many years ago. Those who criticize the annual hunting regulations will do well to ponder the background. Even today disgruntled sportsmen occasionally express the opinion that the Federal regulations are unnecessary and unwarranted, and that the control of hunting for migratory waterfowl should be returned to the states. The pages of history speak to the contrary.

CHAPTER V

Bird Migration

"TO A WATERFOWL"

"Whither, 'midst falling dew,
While glow the heavens with the last steps of day,
Far, through their rosy depths, dost thou pursue
    Thy solitary way?

"Vainly the fowler's eye
Might mark thy distant flight to do thee wrong,
As, darkly seen against the crimson sky,
    Thy figure floats along.

"Seek'st thou the plashy brink
Of weedy lake, or marge of river wide,
Or where the rocking billows rise and sink
    On the chafed ocean side?

"There is a Power whose care
Teaches thy way along that pathless coast,
The desert and illimitable air,
    Lone wandering, but not lost.

"All day thy wings have fann'd,
At that far height, the cold, thin atmosphere,
Yet stoop not, weary, to the welcome land,
    Though the dark night is near.

"And soon that toil shall end;
Soon shalt thou find a summer home and rest,
And scream among thy fellows; reeds shall bend
    Soon o'er thy shelter'd nest.

"Thou'rt gone, the abyss of heaven
Hath swallow'd up thy form; yet, on my heart,
Deeply hath sunk the lesson thou hast given,
    And shall not soon depart.

"He who, from zone to zone,
Guides through the boundless sky thy certain flight,
In the long way that I must tread alone,
    Will lead my steps aright."

—WILLIAM CULLEN BRYANT.

THERE is probably no subject of Nature that has perplexed and intrigued man throughout the centuries as has the movement and flight of birds. It was the mystery of bird migration that led writers in Biblical times to ponder over the comings and goings of hawks, storks, doves, cranes, and swallows. It was the riddle masking the ability of birds to soar smoothly through the air that prodded Orville and Wilbur Wright to risk their necks at Kitty Hawk years ago to prove to themselves and the world that man could also taste the thrill of being lifted and borne away by the restless currents. Perhaps that urge is not so strange. After all, psychologists tell us that this little-understood urge to move about with the seasons is expressed in we humans by the familiar "spring fever" that overtakes so many of us when the leaves begin to break out and the warm breath of spring makes office work a thorough and complete bore. They even hint that youngsters "playing hooky" from school may well have been induced to slip away from their classes in response to that urge to do something about the human emotions that accompany changing seasons.

Various attempts to learn about the movements and habits of birds have been made for generations. One of the earliest records of marking birds for later identification describes how a Roman sportsman took swallows from Volterra to Rome. After the chariot races, the birds were marked with the winning colors and released to fly to their home roosts carrying the news of the winners. The records fail to disclose whether this particular Nature student was also a gambler, but if he was, the ponies should have been very good to him. The first known record of placing a metal band on the leg of a bird involved a common heron captured in Germany in 1710 which is said to have carried several metal rings. Apparently one of these had been attached in Turkey and as such it created a great deal of excitement.

Bird Banding

The first really scientific attempt to mark birds seems to have been made in 1899 by a Danish schoolmaster, one Hans Chr. C. Mortensen, who used aluminum bands for the scientific study of white storks, European teals, starlings, and two or three birds of prey. He can be credited with originating the idea that has since given the scientific world so much useful information.

The work of Mortensen intrigued the interest of many ornithologists in Europe and soon there were banding projects in operation in Germany, Hungary, Holland, England, Scotland, Portugal, Bavaria, Switzerland, Sweden, and other nations. Although Audubon had experimented with bird banding in 1803, the first scientific bird banding in America was by Dr. Paul Bartsch, of the United States National Museum, who banded about 100 young black-crowned night herons in the District of Columbia during the years 1902 and 1903. Other early workers were Dr. Leon J. Cole, P. A. Taverner, and Howard H. Cleaves, who served as secretary of the American Bird Banding Association when it was formed in New York City in 1909. The Linnaean Society of New York was also very much interested and gave financial aid to the bird banding work.

Interest in this new method of bird study grew rapidly and bird banding became such a burden upon the time and resources of these private organizations that in 1920 it was turned over to the Bureau of Biological Survey to be carried on as a part of the official research on the distribution, migration, and other habits of birds. For years bird banding has been one of the important functions of Service personnel, particularly on the wildlife refuges where birds concentrate in large numbers and can be readily taken. The Dominion Wildlife Service of the Canadian Department of Mines and Resources cooperates closely with the Fish and Wildlife Service in this undertaking. There are approximately 1,000 active bird banding cooperators working in the United States and Canada under permits issued by the Fish and Wildlife Service and the Dominion Wildlife Service.

Throughout the years the most extensive files of data pertaining to bird migration that have ever been assembled in the world have been built up and are now housed at the Service's Patuxent Wildlife Research Station at Laurel, Maryland, near Washington, D. C. Scientists from all over the country constantly visit the laboratory to consult these records. More than 5,500,000 birds have been marked with Service bands and the data recorded in the Service's files. About 16 per cent are ducks, geese, and coots. Almost 400,000 recoveries have been recorded, 30 per cent of which were from waterfowl. This disproportionately large percentage of waterfowl recoveries is explainable because few other species may be legally killed. Hunters have been exceedingly cooperative in sending in information about the bands found on the birds taken by them during the shooting seasons. Anyone who sends in a band is notified of the date and the location where it was placed upon its bearer.

The intense interest in this undertaking is well demonstrated by the great numbers of birds that have been banded by organizations and individuals throughout the years. Many people derive much satisfaction from placing the little metal bracelets upon the legs of birds because of the knowledge that by so doing they may aid in determining answers to these age-old mysteries of bird migration. The organizations and individuals that have done the greatest amount of banding with the approximate number of bands used are as follows:

Austin Ornithological Research Station, Cape Cod, Mass.	205,000
E. A. McIlhenny, Louisiana	190,000
William I. Lyon, Illinois	96,000
Dr. Frederick E. Ludwig and C. C. Ludwig, Michigan	82,000
Ben Coffey, Tennessee	67,000
R. J. Fleetwood, Georgia	61,000
Mrs. Amelia R. Laskey, Tennessee	55,000
Ducks Unlimited of Canada	45,000

Many state fish and game departments are using Pittman-Robertson funds to organize banding stations which will further contribute to the knowledge of waterfowl migrations. The outstanding state cooperators, from the standpoint of bands already placed, are Illinois and New York. A number of other states such as Colorado, Maine, Massachusetts, and Minnesota recently have become active in this field. Also, the Fish and Wildlife Service has materially

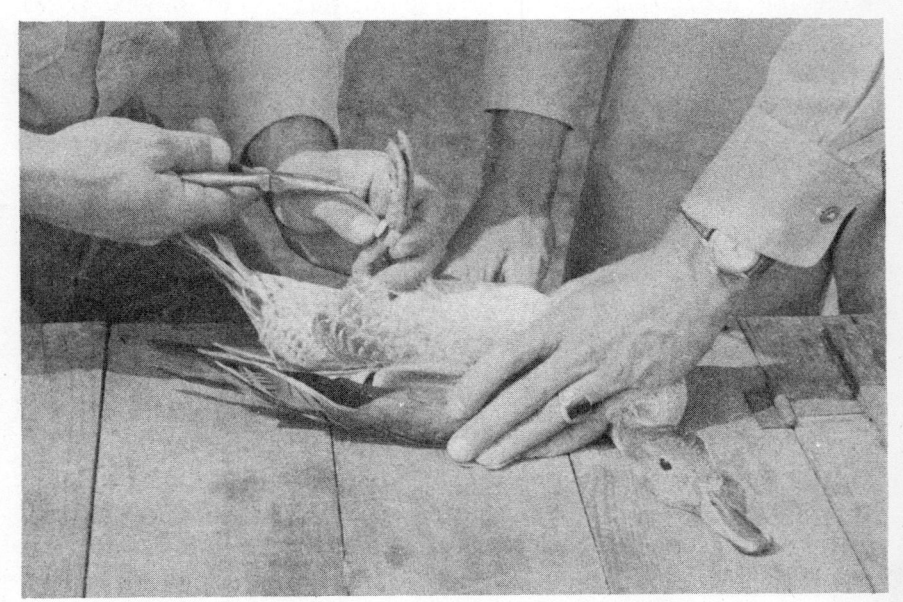

Plate 7. Bird banding reveals many secrets of waterfowl habits.

stepped up its own banding operations, both in the United States and in Canada, as has also the Dominion Wildlife Service.

As a means of effecting proper cooperation and coordination in the bird banding operations, all bands placed on migratory birds in the United States, and most of those in Canada, are furnished to the operators by the Fish and Wildlife Service. Operations by the late Jack Miner at Kingsville, Ontario, chiefly with Canada geese, are the only important exception. The bands are made of aluminum, and range in size from tiny ones for wrens and warblers to those resembling a man's large finger ring, designed to fit the legs of geese, cranes, and pelicans. Each bears a serial number and an inscription reading:

NOTIFY FISH AND WILDLIFE SERVICE,
WASHINGTON, D. C.

Recoveries come from far and near. They are reported by observers who capture birds in other traps on the wintering grounds; they come in from persons who find birds that have been killed by flying into telephone or electric wires, or tall buildings, or monuments; from birds that have perished during long over-water flights and have been washed ashore; from hunters who have killed game species during the hunting season; or from the cooperators who did the original banding and who often recapture the same birds at their stations in the same traps during following seasons. Thousands of bands placed on geese and ducks that were trapped and banded on the wintering grounds in Illinois, Arkansas, and Louisiana have been collected by Service biologists in sub-Arctic regions. Many of these come from Indians and Eskimos in their annual drives on the nesting grounds to capture the birds for food and clothing. Bands have been sent in from such regions as the British Isles, France, Africa, Eastern Asia, practically every country of Central and South America, and from isolated pin-point islands in the South Pacific Ocean.

With the consent of the Canadian Government, the late Jack Miner, one of the original exponents of waterfowl refuges, used his own bands on birds trapped during their welcome stay at his famous sanctuary at Kingsville, Ontario. A God-fearing and religious man, he conceived the idea that he could spread the gospel by quoting Biblical phrases on the bands placed on the legs of his feathered friends. On one side of his metal band he had inscribed: "Write Box 48, Kingsville, Ontario," and on the other side, some Biblical verse such as: "He careth for you. 1 Pet. 5:7."

Jack Miner reported:

In less than a week I had the fowls of the air carrying the word of God and in six months they were delivering it from the sunny side of the Atlantic to the far-off Indians and Esquimaux of Hudson Bay. * * * *

One duck, killed in Louisiana, brought to my home thirty-nine interesting letters of inquiry. Among them was a letter from the Arkansas State Prison, reading as follows:

Plate 8. Bird leg bands. This photograph is actual size.

"My name is ———. My room-mate's name is ————. I am in here for overdraft on a bank; my room-mate, who is sitting at my elbow, is in here for murder. We have a paper here giving an account of a duck killed in Louisiana with a tag on its leg marked 'Have faith in God.' We have looked this up in our Bible; we find that the reference given is correct. We would be pleased to hear from you, to know more about your interesting life with the birds. However, if you do not see fit to write us, we trust you will not be offended at getting a letter from here. We remain, Yours, ——— Arkansas State Prison."

Little did I think when I stamped this verse on the tag that the duck carried away, that the message would ever find its way into a prison cell, and lodge in the heart of a murderer.

During the early days of the banding project, there was no American manufacturer prepared to make the bands, so supplies were ordered from England. On one lot the Biological Survey abbreviation, "Biol. Surv. Wash. D. C." was misspelled to read, "Boil. Surv. Wash. D. C." A Kansas farmer shooting a crow carrying one of these bands is alleged to have reported: "Dear Sirs: I am reporting one crow I shot wearing a metal band numbered 12694. I should report that I followed instructions on the band but am badly disappointed in the result. I Washed, Boiled and Surved, but the durned thing still wasn't fit to eat."

Unusual Records

One of the most dramatic discoveries made entirely through the medium of banding was the location of the winter home of the chimney swift. Although an abundant bird during the summer in eastern North America, the location of its wintering ground remained one of the most intriguing mysteries of bird life. Banding gave the solution to this puzzle. There was great excitement in the bird banding office when a report, dated May 23, 1944, was received from the America Embassy at Lima, Peru, giving the serial numbers of 13 bands turned in by a group of Indians who had killed the birds that bore them at a point on the River Yanayaco, near the boundary between Peru and Columbia east of the Andes Mountains in the upper part of the Amazon basin. Service files showed that these bands had been placed on the swifts in Ontario, Connecticut, Illinois, Tennessee, Alabama, and Georgia. Thus, the mystery of the wintering ground of the chimney swift was solved.

Another important ornithological discovery that can be credited directly to banding is the migration route of the Arctic tern. This species breeds in North America south to Great Slave Lake and to the coast of Massachusetts. After the young are grown, the terns disappear from the American breeding grounds and a few months later may be found in the extreme southern coast of South America and Africa or beyond, even in the Antarctic region, 11,000 miles away. The flight of the terns remained a mystery for many years, but birds banded along the coast of Maine and Labrador in July were found in southern France and western Africa in October and November. One record indicated that a young bird had made a flight of 8,000 or 9,000 miles when it was less than four months old.

Large numbers of birds of various kinds have been banded at the Bear River Refuge in Utah. The vast area surrounding the Great Salt Lake has long posed a serious problem because of the prevalence of botulism, or western duck sickness. For years the Fish and Wildlife Service has maintained a crew at the headquarters of the Bear River Refuge to recover birds in the early stages of botulism poisoning and to treat them until they recover. Then they are liberated. One long-distance record is held by a pintail banded from among the botulism patients on the Bear River Refuge on August 15, 1942. It was recovered 82 days later at Palmyra Island in mid-Pacific, nearly 1,100 miles south of Honolulu. The over-water flight was in excess of 3,000 miles. Another interesting pintail record from the Bear River Marshes came from a bird shot at Toluca, near Mexico City, Mexico, during the fall of the same season in which it had been banded. On the same day at the same spot the hunter who killed this pintail took another which had also been banded during that same season near Dawson, North Dakota. A Bear River pintail also holds the waterfowl old-age record, having been taken near Mexico City thirteen years later. Another banded at Ellinwood, Kansas, was taken more than 3,300 miles to the northwest on the Kobuk River, Alaska, while a fourth banded in Oregon, was killed 2,800 miles to the southeast near Beliz, British Honduras later in the year. Pintails are wide travelers.

A blue-winged teal banded at Ellinwood, Kansas, was recovered in Elia, Camaguey, Cuba, the following autumn. Another banded at the same station was recovered at Corocito, Honduras, while a third banded at Kearney, Nebraska, flew southeast about 2,600 miles to Santa Marta, Columbia. A lesser scaup banded in Colleton County, South Carolina, was recaptured in the Upper Tanana River region in Alaska, while another lesser scaup banded at Lake Athabaska in northern Alberta was recovered at Chanquinola, Panama, two months later.

Banding records disclose that the long-legged, slow-moving great blue herons also get about over the country. One banded at Waseca, in southern Minnesota, was recovered at El Hule, in the State of Oaxaca, southern Mexico. Another one banded at the same station was recovered at Gatun Lake, Panama, and a third banded at Green Bay, Wisconsin, was killed at Point Cormenal, Cuba.

A classic illustration of longevity and of homing instinct is furnished by a mallard duck that carried a Biological Survey band placed on its leg in November, 1927, at a small game refuge near Antioch, Nebraska. This duck for seven years returned and occupied the same nesting site which, curiously enough, was a box on the roof of a barn. She was known to have produced more than 100 ducklings, many of which also were banded. A few of these were recovered, all in the Central Flyway but none in the State of Nebraska where they were hatched. This seems to indicate rather definitely that, while these young ducks kept to their native flyway, they were not particularly interested in returning to the exact location of the mother's nest during subsequent seasons.

Another instance of the native inclination and ability of waterfowl to return to their ancestral homes was well demonstrated by an experiment conducted by the Service a few years ago. Ducks were trapped on Mr. McIlhenney's station at Avery Island in Louisiana, shipped by air, and then released simultaneously in Washington, D. C., on Lake Merritt in Oakland, California, and at several other points in the Atlantic and Pacific Flyways. The following year, all but a few of those birds were retaken from points along the Mississippi Flyway, and a number were recaptured in exactly the same traps on the same identical spots on Avery Island.

Practical Results of Banding

Waterfowl banding is one of the most useful management devices available to the regulatory agencies for obtaining data on which to base hunting regulations. While much has already been

Plate 9. Research has revealed much information on the migratory habits of waterfowl.

learned, the potential value of well-planned projects in strategic production and wintering areas holds great promise for obtaining data which will be of increasing value, particularly when more closely correlated with the annual inventories and the breeding ground surveys. Here are some of the returns to be expected from well-organized, extensive banding operations: a closer approach to the total waterfowl population on the North American continent than it has been possible to obtain to date; a more accurate delineation of the flyways; the value of the various specific breeding grounds in supplying ducks and geese to specific hunting areas; variations in the total hunting pressure from year to year and among the different sections of the country; the over-all value of different species in supplying sport for hunters; losses in population due to hunting, disease, accident, lead poisoning, and predation; sex and age mortality, which is important in evaluating the classes of birds that bear the brunt of the hunting pressure; longevity and the period during which the population of any given year continues to supply sport; and an index of the total annual kill of waterfowl.

These questions are all highly important to management, and for that reason the Service has expanded its banding operations as rapidly as funds would permit. The states also have been encouraged to organize banding projects, the returns of which will be correlated with the efforts of Service biologists. Regulations can be made with greater accuracy as more banding stations are established on the important breeding grounds, the migration stop-overs, and on the wintering areas.

Extensive trapping and banding of Canada geese at the Illinois State Refuge, Horseshoe Lake, Cairo, Illinois, gave a specific answer to the movements of much of the large concentration of geese that annually winter in that vicinity. Robert H. Smith, one of the Service's flyway biologists, at one location near James Bay in Canada, recovered more than 100 Canada goose bands from natives in that vicinity who had taken the geese for food and clothing. A large portion of these were Horseshoe Lake birds. Banding on the Bear River marshes in Utah has demonstrated that Canada geese raised on this Federal refuge are largely residents of the Great Basin country of Utah and eastern Nevada. Sixty-six per cent of the band recoveries of Bear River Canada geese were from within a radius of 100 miles of the breeding grounds in Utah, with smaller percentages in Nevada and California. During the summer of 1947, Service biologists banded 244 Canada Geese at Lake Newell, Alberta, and during that fall,

Plate 10. A redhead proudly wears her band.

recoveries demonstrated that hunters shot at least 49 of these birds during the first season. Thus, analysis of these data shows that the hunting pressure in the fall of 1947 was responsible for taking one out of every five birds during the first year. Recoveries further showed that the Lake Newell geese migrated south through Montana into the Snake River country of Idaho, thence across northwest Nevada to Mono County, California, where many of them winter. There was no evidence that this flight had mixed with geese from Saskatchewan. These records also indicated that Alberta hunters benefited more than American gunners, as 71 per cent of the returns came from that province. California ranked next with 16 per cent, and Idaho third, with 8 per cent. These are just a few examples of the kind of information that banding operations produce.

CHAPTER VI

The Flyway Concept

BANDING studies have established the fact that avian life confines its migratory movements to rather well-defined geographical regions, which we have come to know as "flyways." Each encompasses a vast continental area. Some of the birds may nest and rear their young in the far North and then find winter food and sanctuary in the deep South. Some species range all the way from the sub-Arctic southward to northern and central South America; others from Labrador and Greenland to the Caribbean; and still others may move about within much more restricted bounds.

Flyways should not be considered as travel routes for waterfowl only. Migratory birds of all species seem to follow fairly well the same patterns of movement as ducks and geese. Also, it should be clearly understood that definite lines cannot be drawn to mark the outer boundaries of any particular flyway. There is always overlapping and shifting of certain portions of the various populations between the flyway zones. Water conditions, particularly in the prairie states, influence the migration pattern of waterfowl from year to year. During periods of excessive drouth in the Dakotas and Nebraska there has been noted a corresponding increase in the number of water-loving birds that move through the lake country of Minnesota and Wisconsin. Later, when water conditions improved on the prairies, the birds adjusted their movements accordingly.

From year to year weather conditions play an important part in the migration of waterfowl. A warm, late fall will retard the southern movement, as the birds remain north longer than they do when autumn storms keep them stirring restlessly about, and when freezing weather, ice, and snow place their food supplies out of reach. The results become evident in the timing and intensity of the fall migrations and wintering concentrations. As an example, the aerial survey made by the flyway biologists of the Fish and Wildlife Service during one winter indicated large numbers of birds in Mexico, particularly along the western side from the Gulf of Lower California southward. These same observers, covering the identical areas at a little earlier date during the following year, found a startling decrease in the number of birds as compared with the previous winter. The decline was so marked that they made a second complete airplane flight paralleling the time of the first survey but extending beyond Mexico into Central America. Still the findings were the same.

Why the great difference? Probably because of generally mild weather during the autumn and probably also because southern California, Arizona, and New Mexico were then experiencing one of the worst drouths in history. A great stretch of land extending from the Grasslands country near Fresno, California, southward far into Old Mexico was virtually without water. Tulare and other lakes that normally supplied suitable environment for migrating waterfowl were dry. It seems entirely logical to assume that the birds failed to find conditions to their liking across this broad strip of semi-desert, and returned northward. The inventory of that January showed a heavy increase in wintering waterfowl in the United States and Canadian sections of the Pacific Flyway—in fact, an improvement of more than 50 per cent, but to balance this increase the Mexican portion of the flyway came up with almost a 50 per cent loss. When weighed together, the net result showed far less improvement than the United States portion had indicated.

The migration routes of the individual species that move southward across the continent in the fall each year follow rather well-defined patterns in each flyway but in the spring there is frequently an intermingling of those travel strands as they reach the breeding

THE FLYWAY CONCEPT

marshes. The principal waterfowl breeding range extends roughly from about latitude 40° northward to the Arctic coast, although there is scattered nesting of some species considerably south of this line. The areas in which at least 90 per cent of the waterfowl are produced are indicated by the dark shading on the map reproduced as Figure 1.

Figure 1. Where ninety per cent of the waterfowl are produced.

As the fall movement toward the wintering grounds progresses, the migrating families begin to merge into larger flocks. Black ducks from the St. Lawrence, honkers from the eastern side of Hudson Bay, scaup and canvasbacks from Manitoba, the Northwest Territory or the Mackenzie delta, perhaps even a few redheads from the Bear River Marshes in Utah, will renew acquaintances as they stop to

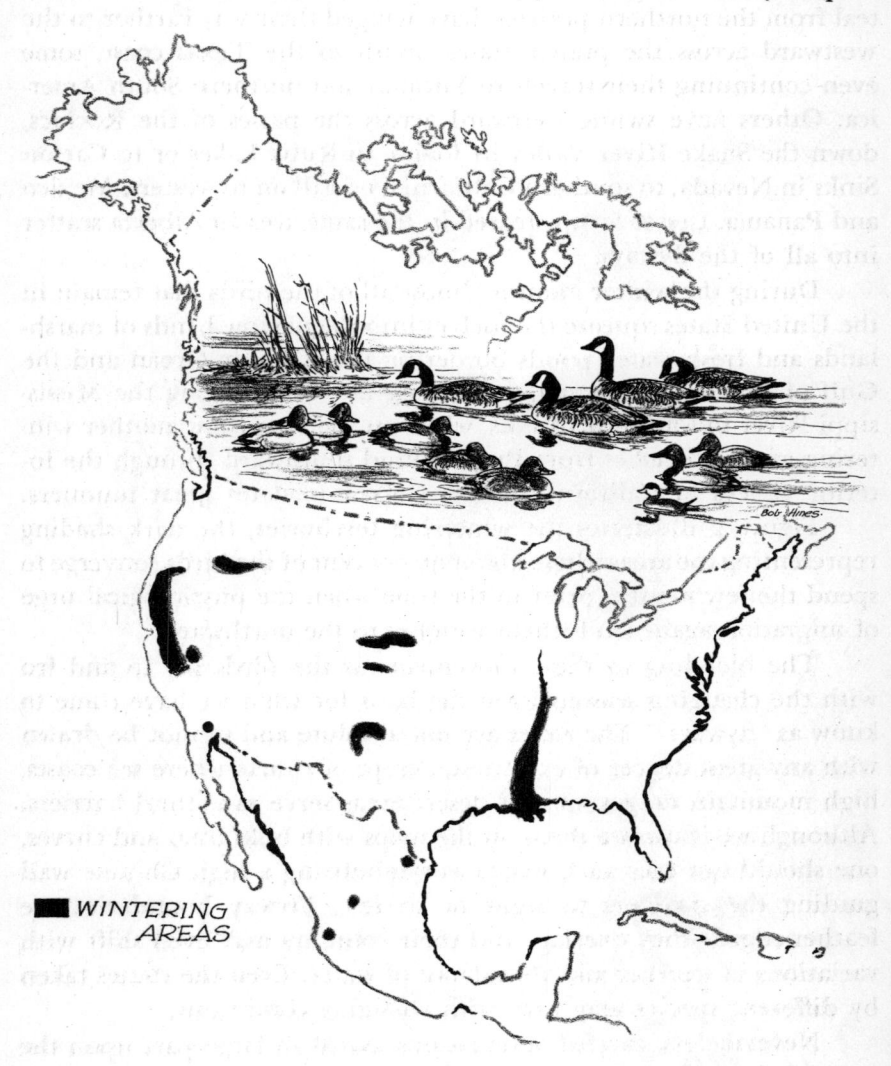

Figure 2. Where ninety-five per cent of the birds winter.

feed along the Jersey coast or the Susquehanna flats on the leisurely journey to Lake Mattamuskeet, Pamlico Sound or, for some of the ducks, to Florida or the West Indies. Other Canada Geese and black ducks raised in the same general localities in the North have taken other routes. Some go down the Ohio and Mississippi to the Gulf. Still other geese, with scaups, redheads, pintails, and blue-winged teal from the northern prairies, have winged their way further to the westward across the prairie states, south to the Texas coast, some even continuing their travels to Yucatan and northern South America. Others have swung westward across the passes of the Rockies, down the Snake River Valley in Idaho, to Ruby Lakes or to Carson Sinks in Nevada, to southern California, or still on to western Mexico and Panama. Lesser scaups reared in the same area in Alberta scatter into all of the flyways.

During the winter months almost all of the birds that remain in the United States squeeze themselves into the narrow bands of marshlands and fresh water ponds bordering the Atlantic Ocean and the Gulf of Mexico with a strip extending northward along the Mississippi River to southern Illinois, while on the west coast another wintering ground reaches from Puget Sound southward through the interior valleys of California. Mexico accommodates great numbers.

Figure 2 illustrates the wintering territories, the dark shading representing the areas where about 95 per cent of the birds converge to spend the few months prior to the time when the physiological urge of migration again sends them winging to the northward.

The blending of these movements as the birds ply to and fro with the changing seasons form the basis for what we have come to know as "flyways." The zones are not absolute and cannot be drawn with any great degree of exactness, except of course where sea coasts, high mountain ranges, or vast desert areas serve as natural barriers. Although we illustrate them on the maps with bold lines and curves, one should not take such marks as symbolizing a high Chinese wall guiding the travelers to right or to left. Flyway boundaries are feather-edged—they overlap, and their contours may even shift with variations in weather and abundance of water. Even the routes taken by different species may vary with changing conditions.

Nevertheless, careful observations, based in large part upon the mass of data obtained by banding, have established migration patterns that can roughly be divided into four zones. These have come to be known as the Atlantic, Mississippi, Central, and Pacific Flyways.

The Atlantic Flyway

The Atlantic Flyway (Figure 3), may roughly be described as encompassing a breeding range extending from Greenland westward to the Mackenzie delta and southeasterly across the prairies of Canada to northern Minnesota and the Great Lakes. The western boundary crosses Lake Erie and northeastern Ohio, thence across northern West Virginia to the Atlantic coast. Here, the westerners mingle with their eastern neighbors, joining one another in concentrations that extend all along the coast from Delaware and Chesapeake Bays southward. Some even proceed to Cuba and the waters of Hispaniola and Puerto Rico.

The spectacular masses of greater snow geese at the Back Bay Migratory Bird Refuge in eastern Virginia, which must be seen to be really appreciated, have flown there with few stops from their nesting sites on bleak Arctic islands and the coast of Greenland. From the same environs have come the American brant, which make the Pea Island and Brigantine Refuges and other areas along the New Jersey, Maryland, Virginia, and North Carolina coasts their cold weather headquarters. The migration routes of these birds represent the eastern boundary of the Atlantic Flyway.

Down across the Great Lakes from the Canadian interior and the far Northland move Canada geese that have nested along the eastern shore of Hudson Bay; black ducks from southern Ontario or the Lake Erie marshes; canvasbacks and redheads from the Delta marshes and the Pas; blue-winged teal that may have nested anywhere from Great Slave Lake to North and South Dakota; and bluebills that may have come from Alaska, the Yukon, and the Northwest Territory. All of these species, and many others, find their way to the Chesapeake and Delaware Bays and the shallow sounds of the Atlantic coast. Strangely enough, a few redheads, banded in the Bear River Marshes of Utah, have been taken on Chesapeake Bay in Maryland.

Oddly, under embarrassing circumstances, I once saw a white-fronted goose killed at Lake Mattamuskeet in North Carolina. Several years ago, an associate and I took Buford Bridwell, a Virginia quail-shooting friend of ours to Mattamuskeet and promised him his

THE FLYWAY CONCEPT 77

Figure 3. The Atlantic Flyway.

Figure 4. The Mississippi Flyway.

first goose. He admitted that he didn't know one duck from another, and in those days canvasbacks, redheads, buffleheads, wood ducks and brant were given complete protection, so we cautioned him to shoot only when given the all clear by Bill Cahoun, the guide. We sat in the blind all day long in nice, clear, sunny weather with nar'y a bird coming within range. Then just before closing time three geese came in from the sun. We crouched with bated breath until Bill said "Now!" Just as we raised he yelled "Brant!" which to all except the Virginia quail hunter meant "don't shoot!" But by this time Buford had forgotten all the species he was supposed not to kill. "Wham" went his old 12 gauge, and down came a "brant," the only bird of the day—and it illegal. The guide was plenty sore. He retrieved the fallen bird in complete and thorough disgust—and then we saw him grinning with relief. Bridwell had killed a white-fronted goose, fully legal but extremely rare along the Atlantic coast. It is the only white-front record for Mattamuskeet.

The black ducks and the Canada geese of the Atlantic Flyway can be separated into two rather general groups. In the Middle and South Atlantic states these two species come from two widely separated regions. The New England area, the Maritime Provinces of Canada, eastern Quebec, Labrador, and Newfoundland represent nesting grounds quite separate from those found in the other region which includes interior Ontario, western Quebec, and the eastern Lake states. The blacks and honkers from the eastern breeding range winter in the more northerly portion of the wintering range while the birds from the interior areas go further south. There is considerable overlapping on the coastal areas from Maryland south, but the interior breeding birds rarely occur on the coast north of Maryland. Indeed, the Atlantic Flyway presents a complicated maze of individual threads of waterfowl movement.

The Mississippi Flyway

The Mississippi Flyway (Figure 4), when projected on a map of North America, looks like a huge funnel located a little east of center. One lip rests on the coast of Baffin Land, and the other on the northwestern coast of Alaska. As the migrants progress southward across Canada into the States, the Missouri and Ohio Rivers pull them together, with the funnel spout leading them down the Mississippi River to the delta, where they spill out westward to the marshlands of Louisiana and Texas. Eastward, they disperse as far as the St. Marks National Migratory Waterfowl Refuge in western Florida.

This flyway is characterized by its large population of mallards that come from widely separated breeding marshes and concentrate by the teeming thousands along the Illinois River bottoms, the Mississippi pools, and in the rice fields and pin-oak swamps of Arkansas. In late autumn I have stood on one of the low bluffs that border the Mississippi River in southern Illinois and have been fascinated and thrilled beyond words watching the distant strands of mallards pour out of the open-water pools. The flocks, in undulating waves, were far out of shotgun range as they passed overhead bound for far-away cornfields.

On another occasion in the dead of a cold winter, I watched a never-to-be-forgotten sight when no less than a quarter of a million ducks, mostly mallards, sat complacently on foot-thick ice at the Chautauqua National Wildlife Refuge, in southern Illinois, black rafts of them spotted here and there over the 4,000-acre man-made lake. The nearest of the sunning and preening masses reluctantly took wing amid great clamor as we walked out toward them on the ice. Then it was revealed that they had kept open a sizeable water hole in spite of the sub-zero weather. So long as they could cooperatively keep the water astir and with the warmth of their bodies prevent it from freezing, and while there still remained corn which the mechanical pickers had dropped in nearby fields, the ducks were quite content to remain.

Wood ducks breed in large numbers in the Mississippi Flyway. It is also characterized by large numbers of ring-necked ducks which breed from Wisconsin northward to eastern Alberta. Lesser scaups and a few other deep water ducks also use this thoroughfare. Canada geese, which nest from the western side of Hudson Bay to Alaska and southward to northern Minnesota and Michigan, winter all the way southward from Wisconsin to the Gulf coast, the bulk of them in southern Illinois and eastern Missouri.

Of particular significance is the fact that although the Mississippi Flyway draws its supply of waterfowl from an enormous territory, the wintering range is sorely restricted. The heavy local winter concentrations lead to assumptions of great abundance—conclusions that may be erroneous. Such a philosophy has often brought on disastrous local slaughter. No less than 120,000 mallards found their way to market from one small rice field area in Arkansas before the Migratory Bird Treaty Act outlawed the sale of waterfowl.

Blue geese that winter mostly along the Louisiana and eastern Texas coasts furnish a good example of the intermingling of migration patterns that characterize the flyways. Nesting in a sub-Arctic environment from the tundras of western Baffin Island to northern Hudson Bay, these fine birds usually take through passage direct to the Gulf of Mexico. Unless forced down by storms, they are seldom seen until they appear in the marshes of Louisiana and Texas. Of late years, increasing numbers have been stopping to the northward, and some have been taken in Michigan, Ohio, and Arkansas. A few have also stopped at the Wheeler National Migratory Waterfowl Refuge on one of the TVA impoundments in northern Alabama. During the spring migration, they move slowly northward, making spectacular and noisy stops in Missouri, Nebraska, Iowa, South Dakota, and Minnesota, before returning to the bleak, inhospitable tundra of the North.

The Central Flyway

The Central Flyway (Figure 5), sweeps in a broad band across North America, the United States portion roughly extending from the Continental Divide to the eastern borders of the Dakotas, Nebraska, Kansas, Oklahoma, and Texas. In the Far North, it reaches from Nome and Point Barrow eastward along the Arctic Coast to about midway in the Northwest Territories. South of the Rio Grande this flyway extends along the eastern half of Mexico and Central America to the waters of northern South America.

This is a flyway of abundance and variety of species when the sloughs and potholes dotting the country from Nebraska to northern Alberta are filled with water and nesting puddle ducks. It is likewise a flyway of scarcity when drouth has passed its withering wand over the prairies. Both the Central and Mississippi Flyways are more subject to population fluctuations than are the other two, chiefly because they are so dependent upon production from the prairie potholes.

Duck hunting in the Dakotas provides many thrills and surprises. On a windy morning one may bring down mallards, shovelers, pintails or widgeons as they move about from one marsh to another;

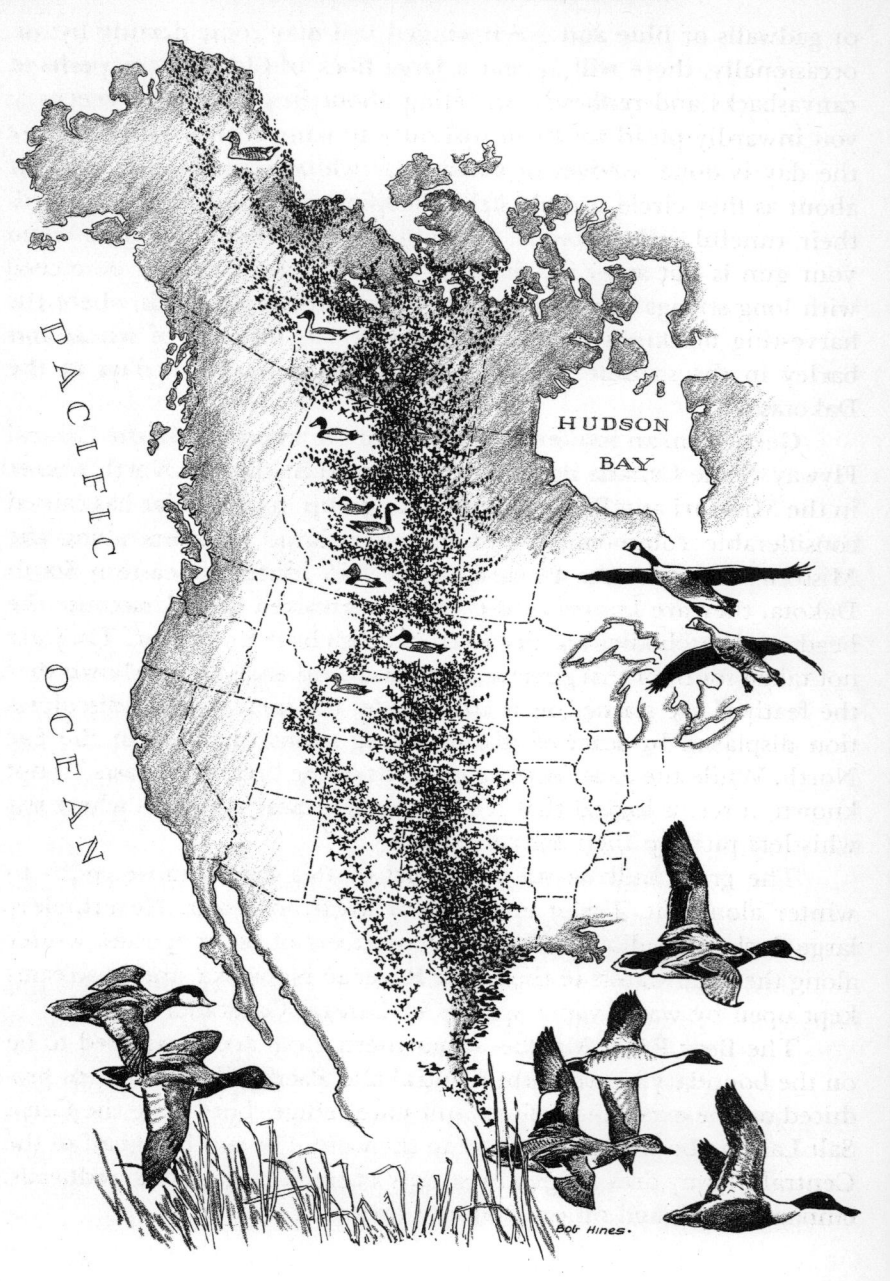

Figure 5. The Central Flyway.

or gadwalls or blue and green-winged teal may come darting by; or, occasionally, there will appear a large flock of bluebills, or perhaps, canvasbacks and redheads, wheeling about in great, wide sweeps as you inwardly plead for them just once to wing within range. Before the day is done, wedges of Canada or white-fronted geese will stir about as they circle stubble fields, keeping your nerves on edge with their tuneful calls. Toward evening—usually about the time when your gun is put away for the day—the Dakota sunset will be etched with long strings of mallards winging toward grain fields where the harvesting machines have scattered generous leavings of wheat and barley in the stubble. I shall never forget a few such days in the Dakotas.

Geese form an important segment of the waterfowl in the Central Flyway. Some Canada honkers which breed in the Far North winter in the Missouri and Platte valleys. One group in particular has caused considerable comment. Killed in rather large numbers along the Missouri River in the Platte-Chamberlain section of eastern South Dakota, they are known as the "golden breasted geese," because the heads and forebodies are distinctly reddish-brown in color. They are not taken south of this general area. Chemical analysis has shown that the feathers are stained with iron oxide, the same natural discoloration displayed by some of the whistling swans that nest in the Far North. While the exact nesting grounds of the "golden breasts" is not known, it seems logical that it is somewhere near the areas where the whistlers pick up their tell-tale stains.

The great bulk of all waterfowl of this flyway move south to winter along the Texan and eastern Mexican coasts. Nevertheless, large flocks of mallards, with lesser numbers of other species, winter along the open waters of the Platte River in Nebraska, and in streams kept open by warm water springs in Central Colorado.

The Bear River Marshes of northern Utah are considered to be on the boundary between the Central and Pacific Flyways. Birds produced on the excellent Federal and state refuges bordering the Great Salt Lake go both to the east and to the west. The southern half of the Central Flyway draws a goodly supply of pintails, gadwalls, redheads, cinnamon teal, and others from this area.

The Pacific Flyway

The United States portion of the Pacific Flyway (Figure 6), extends roughly from the western Rocky Mountains to the Pacific coast. Alaska and western Canada supply the principal breeding range, while western Mexico furnishes wintering habitat for a large segment of the birds of this flyway. Some birds even move on across Central America into northern South America.

The migration routes of this flyway are characterized by great flights of pintails, widgeons, and spoonbills, with smaller numbers of gadwalls and green-winged teal, and a wide variety of geese. It, too, has some peculiar ribbons of movement. Black brant, some of the sea ducks and the diminutive cackling goose leave the nesting areas of the Far North and follow along the British Columbia coast, where there also are offshore flights of pintails, sea ducks and, perhaps, some other waterfowl from the Alaskan Peninsula. The cackling geese that breed in the Yukon delta follow the same general airway, coming inshore near the mouth of the Columbia River and, turning south again, finally wintering at the Tule Lake and Klamath Wildlife Refuges on the Oregon-California line and in the Sacramento Valley in California. A few continue on to the northern part of the San Joaquin Valley.

Many of the ducks and, to a lesser degree, Canada geese, are produced in the United States. Western Montana, eastern Washington and Oregon, and northern Idaho supply large numbers of mallards, gadwall, pintails, shovelers, and teal. Each year, Malheur Lake in Oregon also is producing an increasing number of Canada geese. The Bear River and Ogden Bay Marshes of Utah furnish many of the redheads that are killed in California, and those breeding grounds also supply other game species for many sections of the Far West.

Nevertheless, the bulk of the Pacific Flyway geese and ducks come from central and eastern Alaska, the lower Mackenzie Valley, and other interior points of northern Canada, with lesser numbers from the prairie provinces. Banding records indicate that relatively few of the birds taken in California and Oregon were reared in the southern prairie sections of Canada and that Alaska and western Canada are the chief sources of supply.

I have had the opportunity to fly some 6,000 miles in Alaska, and I wish it were possible for me to describe the vastness of the excellent

THE FLYWAY CONCEPT 85

Figure 6. The Pacific Flyway.

marshlands in the interior portion of the Territory. Over the Tetlin Valley between Northway and the Big Delta, then southward to Anchorage; westward across Lake Illiamna to the Bristol Bay country; northward to Bethel on the Kuskokwim; Hooper Bay on the shores of Bering Sea; eastward to Marshall and Holy Cross on the Yukon; over the marshlands bordering the Innoko and Iditarod rivers; and then eastward toward Fairbanks, an endless expanse of sloughs and marshes unfolded ahead and below us. For mile upon mile upon mile, we saw the same kind of environment—clear, pure, shallow pools, bordered with green and brownish vegetation and connected by narrow guts that meandered to the larger channels that eventually emptied into sluggish rivers flowing westward to the sea. Dropping down to two or three hundred feet elevation, we could see broods of Canada, white-fronted, and cackling geese, baldpates, scaups, pintails, golden-eyes, and canvasbacks, with old squaws and other sea ducks predominant. Flocks of little brown cranes circled below us, while standing out in bold relief were pairs of nesting swans with their grayish cygnets scurrying to cover as we roared overhead.

As we maneuvered between the snow-tipped mountain ranges on aerial trails that often led us over canyon-like crevices staring up from age-old glaciers, I gained some very distinct impressions of Alaska. My pilot, Service Regional Director Clarence Rhode, who is tops among the bush pilots in the Territory, seemed completely unconcerned and blasé about the whole trip; while for the life of me I wouldn't have said so then, I will now admit that mingled with the fascination of the kaleidoscopic landscape rolling out beneath us was a certain amount of worry on my part as to where our old war surplus Grumman Goose might set herself down if her motors should fail. Now that my feet are again firmly planted on terra firma, my most vivid impression of the trip was the immensity of the flatland marshes of interior and western Alaska. I have since roughly estimated that these unspoiled waterfowl production grounds equal an area the size of New York, Massachusetts, Pennsylvania, and Ohio combined. Fortunately, they will probably always remain unspoiled. Would that as much could be said for the rapidly disappearing Pacific Flyway wintering grounds which are just as important to those birds as are the nesting sites.

Figure 7. State grouping for management purposes of the four flyways.

Management by Flyways

As the Service has been able to add to the store of knowledge of waterfowl behavior, it has been possible to gradually move toward the adoption of regulations which conform to the conditions in the general regions that have come to be known as flyways. The Atlantic Flyway is preponderantly a black duck, scaup, baldpate, pintail, and Canada goose route; the Mississippi, a Canada goose, mallard, and ring-neck highway; the Central, a mallard, pintail, teal, and white fronted goose thoroughfare; while the Pacific is noted for pintails, mallards, widgeons, shovellers, teal, and its great abundance of several species of geese. Some of these regions are so situated that they will always have a greater number of certain kinds of birds than others. The Central and Mississippi Flyways will continue to be quickly influenced by changing water conditions on the breeding grounds. The Pacific and Atlantic Flyways situation will become more and more critical as wintering habitat recedes due to the growing demands of agriculture, industry, and commerce.

From some quarters there has been strenuous objection to waterfowl management by flyways because duck hunters in one region insist that their fellow sportsmen in other sections be given no more liberal privileges than they, themselves, enjoy. It is difficult to understand this type of reasoning since they accept the fact that their own states regulate resident game on the basis of local abundance or scarcity. Many waterfowl hunters seem to have developed a different psychology, undoubtedly because ducks and geese are migratory, and some folks do not yet understand that migratory species can and should be managed according to the principle of supply and demand.

In order to avoid the complications that would arise if the Service attempted to establish hunting regulations based only upon the migration routes of the birds, particularly where the boundaries of the flyways bisect portions of different states, the states are grouped into those flyways which claim the bulk of the waterfowl found within their borders. As an example, birds produced in Utah, particularly on the Bear River Marshes, move into both the Pacific and the Central Flyways, but since the majority of them use the Pacific, the state of Utah is included in that flyway. Similarly, in the Dakotas, the western boundary of the Mississippi Flyway follows a diagonal trail across these two states. Since, however, it appears that most of the ducks nesting in or traveling across them belong to the Central Flyway, these states have been included in that group. Figure 7 illustrates the grouping of the states that for management purposes are included in the four flyways. This has been done with the full knowledge that some of the ducks and geese will refuse to cooperate by remaining within the boundaries shown on these maps.

Regulation by flyways probably will be increasingly accurate as more definite information is obtained through banding operations and the year-long observations by biologists of the Fish and Wildlife Service, the state game departments, and interested organizations. As the basic data become better known, there will be a growing public acceptance of the fundamental soundness of managing the waterfowl resource on a flyway basis.

CHAPTER VII

Keeping Up With the Waterfowl

*O*NE OF THE challenges most frequently aimed at us during public discussions on waterfowl management sounds something like this: "How in the world can anyone estimate the number of ducks and geese there are in the country? They breed from the Arctic Circle to Nebraska, and they winter from Alaska to South America, so how come you try to tell us you know how many ducks there are? You can't do it, and you should admit it and stop fooling the public by making them think you can."

To which we candidly admit that we don't know the exact number of ducks and geese and other waterfowl on the continent. Nevertheless, when the time comes around each year to make the regulations, the Service has a great mass of reliable information that indicates whether there are *more* birds or *fewer* birds than there were the previous season. Even if the exact number of millions is not known, there is not much doubt as to whether the trend is up or down and about how much. The finest collection of data that any nation has ever assembled on any wildlife resource is available to us as background for vital decisions. There is nothing in wildlife management anywhere in the world that can even begin to compare with it. And that is as it must be, because no other heavily-populated nation has ever attempted to manage its wildlife resources as have we Americans. We try to do it in such a fashion that it is available for all, rich and poor alike. In Europe, game is still abundant in many places but, there, wildlife is confined to large private estates or to public preserves where hunting is severely restricted and limited. Nowhere else in the world is there the "come one, come all" philosophy such as ours. Therefore, it is imperative that there be abundant factual information available if we are to manage successfully a resource that ranges over the entire continent.

Keeping up with the waterfowl is a year-round job. The Federal refuges furnish a valuable source of information. These 200 units, scattered along the migration routes, are staffed with well-trained, permanent managers who regularly supply information, not only pertaining to the refuges but also to the surrounding country. Since about one-fifth of the continental waterfowl population uses the refuges at various times during the year, these observations provide a constant barometer of the waterfowl situation.

Another group of experts whose job it is to keep up with conditions and populations throughout the year is that all too inadequate staff of Service personnel known as Game Management Agents who make waterfowl management their chief interest in life. They are the Federal law-enforcement officers. Although numbering only about 80, they are in constant contact with the 2,500 state game enforcement personnel so that their knowledge is of greater importance than their small number would indicate. They are all experts in their chosen field and their reports of year-round conditions are invaluable.

And then there is a group of volunteer observers located in all of the important waterfowl concentration areas who have been submitting reports for many, many years. These volunteers have good knowledge of waterfowl, are accurate, and have no bias. Many of them have been giving their time, energy and funds continuously for many years. Their comparisons of abundance and scarcity in all parts of the country furnish a valuable source of information. If any evidence of prejudice or exaggeration shows up in the reports—either on the side of optimism or pessimism, it can be checked by comparison with information from other sources. There are always some people who can see things as they want to see them, regardless of the facts, but those folks can soon be detected and eliminated from the list.

I recall a case in point which occurred during one critical fall when we in the Service were holding our collective breaths to see whether we had done the right thing by permitting a short season with small bag limits, instead of following the advice of thousands of people throughout the country who felt that waterfowl hunting should have been brought to a complete halt. My attention was called to a story which made the sports section headlines in a Lake state

Plate 11. "A Pintail Drake and Hen" coming in for a landing were pictured in the 1938 stamp. This was drawn by Roland Clark. Sale 1,002,715.

paper. It was authored by one of that small percentage of outdoor writers who never overlook an opportunity to take a pot-shot at the Service. It told of an airplane flight this self-styled expert had just made over several lakes in his vicinity. He wrote that he had seen literally millions of ducks—in fact the water was black with them, from which he concluded and loudly proclaimed that those "blankety-blank bureaucrats" who had imposed the outrageously short season that year upon the poor, down-trodden duck hunter were either stupid, incompetent, or dishonest—or more likely, all three.

It made quite a yarn, and many sincere readers believed it. The glowing report of a huge early flight of ducks sounded good and the local Service game management agent immediately rechecked the situation. He hired a plane and flew all over that part of the state. He found the birds, several thousands, rather than millions, but instead of being ducks they were coots, the early flight that always comes down through that section. Moreover, even the coots were not nearly as numerous as they had been the year before. Hardly a duck was to be found. That particular outdoor writer is not on the Service's list of volunteer observers.

Much reliance is placed on the findings of a small but highly-trained staff of Service biologists who spend their entire time in keeping up with the waterfowl. These fellows are important cogs in the vast machine that each year determines what the regulations should be. They are known as "Flyway Biologists" and can best be described as earnest, enthusiastic young scientists who never stay at home. At least, that is what their families say, and it comes awfully close to being the truth. They spend their days following the ducks and the geese—into the Far North in the summer, and southward in the winter. They live ducks, they talk ducks, they think ducks, they dream ducks, they even mumble in their sleep about them—and little wonder!

Did you ever try to count a brood of young ducks? Did you ever sneak through the marsh, come upon a wary old Susie, and attempt to get an accurate estimate of the number of young hiding or dashing and splashing through the rushes? To be sure, if the old girl takes her little flock out into the open and swims down to the end of the pond, you can get a pretty fair sampling of the number of downy youngsters, but if they follow their usual instinct and scurry to cover, then it is a different situation. That sort of observation is one small part of the job of the flyway biologist.

In the final analysis, the hunting regulations must be made on the basis of the current situation on the breeding grounds. It is obviously impossible for anyone to sit down in mid-winter and determine off-hand what the waterfowl hunting regulations should be for the following season. That determination must be based upon information obtained each year in the Northland by trained biologists who observe the approximate success of the spring hatch. If the birds find suitable conditions with ample water when they go north in the spring to nest and make their summer homes, they will bring off large broods. Then the hunting regulations for those flyways can be more liberal than in the years when the birds find the sloughs dry and weed-grown, the potholes mud-caked, and when nesting locations are few and far between. The success or failure of the nesting season is the key to the kind of regulations issued in the fall. The more complete the appraisal, the more accurate the regulations will be.

Transect Sampling

The success or failure of the breeding season can best be measured in terms of trends, and these must be obtained by sampling. Whenever anyone comes out with a figure claimed to be the total size of populations during the breeding season, it would be well to look a bit carefully, because in the light of present knowledge it is impossible for anyone to make much more than a wild guess at that time of the year. The nesting area is too vast. In earlier days the Service relied largely for its data upon the observations of trained biologists who had covered the same areas successively during a period of many years. These men, traveling by automobile, airplane, and canoe, kept records based upon the observations and impressions they received in the field and the local key observers they interviewed.

During recent years biologists of the Fish and Wildlife Service and cooperating agencies have developed a technique that holds great promise for obtaining permanent comparative records of conditions on identical breeding ground areas year after year. Random lines of travel have now been established that will be followed between well-defined points projected over wide areas throughout the breeding ranges. These same routes are followed and observed during successive years, thus providing accurate comparative data.

For instance, in North Dakota there has been charted a total of nearly 1,000 miles of travel where observations are made every year over the same, identical route. If the observer is using an automobile, he travels the backroads course and records carefully key mileage points on the speedometer of his automobile. He then travels along a four-mile stretch from each key point, noting the number and size of the potholes in an area one-eighth of a mile on each side of the road. The two eighth-mile-wide strips, four miles in length, equal one square mile, or 640 acres. In that area he has counted the number of potholes and small lakes, recorded the approximate acreage of each, and with a pair of field glasses and a certain amount of trailing through the marsh, he has estimated the number of breeding pairs of birds. By repeated trips over this same route throughout the summer, the observer is able to determine the fluctuation of the water and predict whether the potholes will hold up long enough to carry the birds through the nesting season and the subsequent flightless moult stage. He knows pretty well the production of each pond in that test area.

Wherever possible, the same observer travels the same route the following year, making identical observations. If, however, he is unable to return, he has left behind a volume of notes so detailed and accurate that another biologist can take over. Thus, both quantative and qualitative data are being accumulated to show the changing conditions on these sample areas of the breeding grounds.

In addition to the observations made from automobiles along the back roads, a similar technique is used by biologists who cover much wider areas in airplanes. Transects are again laid out on aerial charts and the pilot, with an experienced observer, flies at low level over the route, noting the numbers of potholes and their extent, and estimating the number of birds of different species. These observations are continued over the same courses year after year.

Conditions on the Nesting Grounds

A person looking at a map of the major breeding areas of the North might well wonder how anyone could possibly cover enough of that vast country to obtain even a vague idea of the total waterfowl population. Admittedly, the problem is difficult, but it is simplified by the fact that not all of this vast expanse produces large numbers of ducks, particularly of the game species.

The Northland may be considered as being divided into four major ecological categories. The first is the tundra region extending from the line of trees northward to the Arctic Ocean. In this section there is always an abundance of water, but there is little or no soil to produce the food and cover plants that are needed by ducks. Tundra conditions denote short growing seasons and often cold and disagreeable temperatures with high waters and heavy winds. Such conditions are not conducive to sustained waterfowl production except for a few of the species such as the snow geese, the sea brant, and some of the sea ducks. Yet, the tundra country covers a vast area, and its total production at times is large. Swans, white-fronted geese, pintails, mallards, green-winged teal, and a few others find suitable nest-

KEEPING UP WITH THE WATERFOWL

Plate 12. Light airplanes flown by pilot-biologists are becoming increasingly useful in observing summer breeding conditions and winter concentrations of waterfowl, as well as apprehending violators of the regulations.

ing sites scattered widely over this great expanse of land and water. Nesting success varies greatly with climatic conditions from year to year.

Then there is the region of forest lands with much water but few breeding ducks. As with the tundra section, the soil in such areas is quite shallow. Many of the lakes, both large and small, are bordered with rocky shores unproductive of choice waterfowl foods. Water alone does not raise ducks. There must also be suitable vegetation, ample food, and the right kind of conditions for the production of broods of young. Lakes rimmed with stony, barren shores in the heavily forested areas produce few waterfowl.

Other than the tundra and the forested areas, there are in Canada the aspen park lands and the prairies. The park lands produce some birds, but by and large, the heaviest production of waterfowl comes from the open grassy sections such as we find in our own prairie states—the Dakotas, Nebraska, eastern Montana, and western Minnesota. Here, relatively long-growing seasons and rich soils that produce lush vegetation for food and cover, shallow water, and scarcity of aquatic predators such as turtles and great northern pike, combine to produce the ideal environment for waterfowl. While some waterfowl are reared over much of the vast Northland, the bulk of the birds are produced in areas that are reasonably compressed and not nearly so extensive as the map would indicate. What is found on these top production marshlands in Canada and the United States is a good index to the fall flight of ducks, particularly in those states east of the Rocky Mountains.

Summer observations of these critical areas have been expanded as rapidly as funds have become available. Many agencies and organizations cooperate with the Fish and Wildlife Service personnel assigned to these summer surveys. The Dominion Department of Mines and Resources and the game branches of the provincial governments contribute much to the breeding ground studies. "Ducks Unlimited" make their data available. The Wildlife Management Institute, which maintains a cooperative research station at Delta, Manitoba, contributes valuable time, aid, and personnel. The Illinois Natural History Survey has sent its top biologists to Canada. The Winous Point Gun Club of Port Clinton, Ohio, has done the same. Other workers have been supplied by the state game departments of Illinois, Massachusetts, and Wisconsin. A staff of trained Service biologists, working with the River Basin study program of the Fish and Wildlife Service in cooperation with Canadian officials, has surveyed the entire Lake Manitoba, Winnepegosis basin, Cedar Lake, and lower Saskatchewan delta for the Provincial Government of Canada.

These experts—the best that can be obtained in the entire United States and Canada—each year spend days on end checking the waterfowl populations and conditions so that the regulations may be the most realistic that is humanly possible. Service airplanes carry biologists thousands of miles, making criss-cross, low-level observations over the breeding areas. Ground crews cover additional thousands of miles by automobile, canoe, and on foot. In some seasons these combined travels total more than 100,000 miles.

Although the prairie provinces of Canada receive major attention, other portions of the continent are not neglected. The Atlantic Flyway observers move into the river valleys and the lakes of the maritime provinces of Nova Scotia and New Brunswick, then travel along the rocky shores of Newfoundland, the bleak coast of Labrador, and on north to Baffin and Ellesmere Islands. Some biologists push forward into the bush country of the Northwest Territories and the Mackenzie delta with its numerous lakes, marshes, and muskegs. Other observations come from the mountain valleys of British

Columbia dotted with picturesque lakes and with ponds perched on steppe-like benchlands, and from the vast nesting grounds of the Yukon, the Kuskokwim, the Innoko, the Tanana, and the other rivers of Alaska. With modern airplanes and the best facilities that can be provided, the observers spare no effort to get the best information that is humanly possible.

Today, the movements of North American waterfowl are better known than are those of any other continent. Research dating back almost 70 years has revealed the flight habits of the birds, while current studies indicate the comparative abundance or scarcity each year. Although all of the questions surrounding this complex problem cannot be answered with complete accuracy, it is an undisputed fact that no other countries have ever had at hand the mass of data and the advice of such expert groups as the United States and Canadian Governments use in the promulgation of the annual waterfowl hunting regulations.

Winter Inventory

The winter inventory is designed to reveal the status of the waterfowl that have survived the natural vicissitudes of summer and of the fall hunting season. Conducted in January, it gives an index to the relative number of breeding birds that will go back to the marshes and potholes of the North within the next few months. A cattle rancher must know how many cows he has in his breeding herd if he expects to stay in business very long. Similarly, a successful wildlife administrator must have a good idea of the breeding stocks that are available for the production of a new crop of wildlife. The annual waterfowl inventory, taken at a time when the birds are compressed into their winter quarters, supplies that information.

Organized through the regional offices of the Fish and Wildlife Service, the survey covers the entire United States. Flyway biologists and game management agents organize the study. The state game departments give liberal assistance. The U. S. Air Force and the air services of the Navy and Coast Guard, personnel from the Soil Conservation Service, the Forest Service, the National Park Service, and thousands of individual volunteers join in the bill-counting undertaking. Between 2,000 and 2,500 people throughout the country participate.

Actual counts of birds on typical areas are made wherever possible. The findings of the ground crews are checked with those of observers covering the same areas with aircraft. Many summaries are based upon aerial photographs. It is surprising how pictures taken from airplanes over open waters can reveal the actual number of birds, particularly when enlarged or examined under a magnifying glass.

By keeping detailed records of the areas covered and then estimating the percentage of the total known wintering habitat that has been examined during the period, it is possible to arrive at a reasonably accurate picture of the overall waterfowl population. Comparison with the previous year's coverage, after making due allowance for improvements in technique, supplies the answer to trends of abundance from year to year. It is gratifying to note that since the first inventory was made in 1934, fluctuations in waterfowl numbers as determined by this system of checking the trends have been substantiated by the field experiences of the waterfowl hunters.

Yet, observations made in the United States provide only a partial answer. An important portion of the continental population of birds normally goes south into Mexico, Central America, and even northern South America. Some winter in the Caribbean, in Cuba, Hispaniola. Some go to Puerto Rico and the Virgin Islands. Others have remained in Canada, particularly along the sheltered bays of British Columbia and in the Gulf of St. Lawrence. It would be a mistake to assume that the returns from the winter surveys in the United States alone portray the true status of waterfowl abundance. The Service, therefore, conducts midwinter studies as far south as Central America and the Caribbean, and obtains data from Canadian and Mexican officials before reaching final conclusions on the results of the winter inventories.

The Take by Hunters

The extensive and intensive surveys conducted throughout the breeding ranges indicate what the fall flight will be. The winter inventory reveals the relative abundance or scarcity of birds at the end of the gunning season. Yet neither of these supply the answer to another vital question of management—the take by hunters and the losses due to other causes. It is difficult to obtain factual information of the extent of decimation by predators, disease, lead poisoning, and poaching, although such studies are now being conducted by

Service personnel, augmented by research projects of several state game departments with the aid of Pittman-Robertson funds.

It should be a simple matter to determine the legal harvest by the duck hunters, but it is not. The Fish and Wildlife Service has tried many approaches. For two years it urged waterfowl hunters to report their season kills, the estimated cripple losses, and whether

Plate 13. Lynn Bogue Hunt, noted wildlife artist, submitted the winning drawing for the 1939 duck stamp, which he called "A Male and Female Green-Winged Teal." Five others descend for a landing in the background. Sale 1,111,561.

they had observed more or less birds than during the previous season. All of the outdoor magazines cooperated whole-heartedly. State fish and game bulletins carried the plea. The Outdoor Writers Association circulated the request through the daily press. Everyone was urged to fill out a simple score card and send it to the Service for analysis.

The result was surprising. Surprising because it showed such a lack of interest among waterfowl hunters, even though they knew they could thus help their official agencies do a better job of management. More than 2,000,000 duck stamps were sold one year, but there were less than 3,500 score card returns. The plan was attempted the next year, and again the magazines cooperated whole-heartedly. They sacrificed advertising space worth thousands of dollars to print the form. Again, the Outdoor Writers Association and the state magazines broadcast the appeal for information and again the duck hunters failed to turn a hand. With 1,750,000 duck stamps sold that year, this combined effort yielded the insignificant total of 2,000 score cards.

Convinced that the average duck hunter cannot be relied upon to submit his observations (with the exception, of course, of those remarks that have to do with the inadequacy or unreasonableness of the hunting regulations), the Fish and Wildlife Service launched upon a new approach—a modified Gallup poll. Although public opinion polls have occasionally received severe jolts following national elections, the new duck hunter poll seems to hold considerable promise.

Carried on by game management agents, leaders of the cooperative research units, and others in the Service, random telephone calls are made until several thousand purchasers of duck stamps throughout the country have been contacted. Questions asked are simple—the number of ducks and geese killed during the season, the estimated cripple loss, the number of days hunted, the average daily bag, etc.

Some interesting results come from the random sampling by telephone. We have learned that about 13½ per cent of duck stamp purchasers do not hunt. This group includes the philatelists and the smaller proportion of people who plan to go afield but fail to do so. To offset this 13½ per cent of duck stamp purchasers who do not hunt, there are those under 16 years of age who need no stamps, and also those who always like to "take a chance" of hunting without the stamp. It thus seems that the stamp sales represent a good estimate of the number of shooters. Also the sampling revealed that the average U. S. hunter killed less than two ducks per day when the legal bag limit was four birds.

Estimating the numbers of wild things in the forests and fields cannot be done as easily as the rancher counts his livestock or the poultry grower his laying hens. The vagaries of nature often change situations overnight. This is particularly true with waterfowl whose production depends upon weather, water, and food provided over the immense areas of the United States, Canada, and Alaska. Yet, the vast amount of data secured through tireless and painstaking effort, studied and checked throughout the entire year, provide an unbiased and solid foundation for the promulgation of the annual hunting regulations. Certainly, it can never be truthfully said that the regulations submitted to the President for his approval each year represent the thinking of a few swivel-chair biologists who sit in Washington and attempt to prescribe the length of season and the size of bag limit based upon their own personal desires and prejudices.

CHAPTER VIII

The Regulations

THE MIGRATORY Bird Treaty Act completely prohibits the hunting of all birds included within the terms of the treaty with Great Britain *except* as permitted by regulations adopted by the Secretary of the Interior and approved by the President of the United States. Thus, the regulations represent the device by which a share of the annual supply of birds may be taken—always within the fundamental philosophy of the law that the welfare of the birds must be considered first and the desires of the hunters second. It provides that "zones of temperature, distribution, abundance, economic value, breeding habits and times and lines of migratory flight" shall be considered in any regulations prescribed for the taking of migratory birds. This provision contemplates that hunting may be permitted in accordance with the relative abundance of the birds in different sections of the country.

Regulations which insure that the hunting privileges will be distributed on a fair and equitable basis throughout the length and breadth of the United States are difficult to prescribe, and the determination of what they should be is—I might well observe—a most thankless task, particularly when added restrictions are in order.

As agricultural and other economic demands increase, as habitat continues to shrink, as the number of hunters mounts, as better roads, more leisure, and new machines of destruction bring new problems, the task of regulating migratory waterfowl hunting will become more burdensome and, I fear, even more thankless. Heaven forbid!

The most equitable means of distributing the allowable take is to expand or restrict the length of the open shooting period. When a more liberal season can be granted, a greater spread between the opening and closing dates insures a better chance that ducks and geese will pass through every portion of every state during the interim. Of course, that does not mean that each section will enjoy good shooting for the full time. That would be impossible. If that were to occur in every local spot and if any reasonable proportion of the hunters were able to secure their bag limits each day, there would be little left for the next year.

For many years, the open waterfowl hunting season was, in general, prescribed by latitudinal zones extending across the United States. At one period, four zones were adopted, and later, three. The northern zone usually opened in the latter part of September or early October, the central zone about the middle of October or later, and the southern, about the middle of November. All states in each group were permitted a specified number of shooting days beginning on the uniform opening date for that particular zone. The length of the seasons depended upon the supply of waterfowl and were variously set at 30, 45, 60, 70, and 80 days. The state fish and game commissions were generally given the choice of placing their states in any one of the three zones. Those in the north normally chose the earliest season, those in the south, the latest, and in the interior it was not uncommon for states to alternate from one to another, year by year. Under this system, duck hunters along the Atlantic coast were given the same number of shooting days and the same bag limits that were permitted in the Pacific Flyway, although the abundance of waterfowl and the number of hunters were often vastly different.

In more recent years, waterfowl hunting regulations have been promulgated to an increasing degree upon the conditions found in each flyway. Factors of abundance and kill in the four flyways have been used to determine the number of hunting days and the bag limits to be granted in each. Specifically, the Atlantic and Mississippi Flyway states have been held to seasons of 30 days of continuous hunting, or two split seasons of 12 days each, at the same time the Central Flyway was granted 35 days, or two split seasons of 14 days each, and the Pacific 40 days, or two split seasons of 17 days each. Likewise, the number of ducks permitted in the daily bag in the two eastern flyways has been held as low as four at the same time that five birds were permitted in the two western flyways. Similarly, the goose limit in the Atlantic Flyway has been held to one Canada goose, while in the Mississippi it has been two, and on the West coast, two Canada geese plus three of other species.

Such differences in the duck hunting privileges between the Atlantic coast and the Pacific states have naturally brought criticisms from some quarters that the Service has been partial to certain sections of the United States and unfairly discriminatory against others. Many people feel that a species that is migratory should be so regulated that people in all sections of the United States will get equal privileges. Complaints of alleged discrimination of this kind usually disappear when the facts upon which the hunting rules are based become better understood.

Let us lay the factors that determined the regulations for that particular year out on the table and take a good look at them. Waterfowl abundance as determined by the annual winter inventory provides the best index of the relative regional abundance of the species subjected to hunting. Analysis of the data secured during the January inventory immediately preceding that year's regulations showed that the Pacific Flyway had 37 per cent, the Central 23 per cent, the Mississippi 25 per cent, and the Atlantic 15 per cent of the continental supply of birds.

The duck stamp sales which indicate the relative number of hunters showed that for the same period the Pacific Flyway had 20 per cent, the Central 25 per cent, the Mississippi 44 per cent, and the Atlantic 11 per cent.

The harvest by hunters during the previous season as indicated by the hunters' score cards and the sampling by Service personnel and cooperators, showed that the Pacific Flyway took 25 per cent, the Central 19 per cent, the Mississippi 46 per cent, and the Atlantic 10 per cent of the birds.

The two western flyways, therefore, had 60 per cent of the birds, 45 per cent of the hunters, and accounted for 44 per cent of

Plate 14. Speckle-Bellies against the sky.

the previous year's kill. The two eastern flyways, on the other hand, had 40 per cent of the waterfowl, 55 per cent of the hunters, with 56 per cent of the kill. It was on the basis of this information, coupled with late season data of breeding success on the areas that supply birds for the different sections, that the two eastern flyways were given shorter seasons and smaller bag limits than the two western. It was also on the basis of these data that the Pacific Flyway was given the most liberal hunting regulations of any portion of the United States. On the national average, those states had more in birds, fewer in hunters, and one-fourth of the kill.

Progressive states have, for many years, regulated the seasons for upland game hunting on the basis of annual abundance or scarcity. Some counties may be open to shooting while others are completely closed. Longer seasons and larger bag limits may be permitted in some section than in others. Management on the basis of actual field conditions is fundamentally sound. Managing waterfowl by flyways will become a more successful practice as additional information and data are secured through improved techniques and wider coverage. We have found that when those who object to the management of waterfowl by flyways analyze the data and understand the logic behind the action they usually agree with the validity of the procedure.

How the Regulations are Made

The Service has long sought advice in the making of the waterfowl regulations and has called upon the executive directors of the 48 state fish and game departments to serve as its unofficial advisory board. These men hold responsible positions and know the feelings and the sentiments of the people in their own states. They share responsibility for the regulations they recommend because they and their staffs play an important part in enforcement. Naturally, their recommendations are much more beneficial to the Fish and Wildlife Service than are those coming from spokesmen for sportsmen's groups who need assume no responsibility for the things they recommend.

Conferences to discuss waterfowl problems are held with state fish and game officials frequently throughout the year. Service personnel attend the regional meetings of the state administrators whenever these are held and, in addition, waterfowl discussions are frequently scheduled during the meeting of the International Association of Game, Fish and Conservation Commissioners. The public meetings held throughout the country each year give further opportunity for discussions among state and Service personnel of the joint problems involved in this difficult task of managing the nation's waterfowl resource.

Another advisory board known as the National Waterfowl Committee assists the Service in determining waterfowl regulations and management policies. It consists of representatives of the International Association of Game, Fish and Conservation Commissioners, the Wildlife Management Institute, the National Wildlife Federation, the Izaak Walton League of America, the National Audubon Society, the Wildlife Society, the Friends of the Land, and the Outdoor Writers Association of America. Ducks Unlimited was an early member but soon resigned. This group meets with Service officials three or four times a year, is kept posted on current problems, and is invited to submit recommendations on the over-all waterfowl management program.

For those who feel that the Fish and Wildlife Service does not receive advice, I would like to assure them that this is not the case. Likewise, I should record that it would be utterly impossible for us to follow all of the conflicting and confusing guidance proffered from all quarters. Advice, while much appreciated, cannot be con-

Plate 15. "A Pair of Black Ducks," the favorite of East Coast hunters, featured the 1940 stamp. The seventh in the series was drawn by Francis L. Jaques. Sale 1,260,810.

sidered as much more than well-intentioned counsel. Chief reliance must always be placed upon the continuing observations of the Service field staff, the trained workers of the Canadian Government, and the states and cooperating organizations.

After the regulations are approved by the President and published in the Federal Register, they become law. Then the trouble starts. Although they are accepted with little question by, I would say, 98 per cent of the wildfowling fraternity, there is always that small number of individuals throughout the country who are certain that they know more about the situation than any handful of "seat-warming bureaucrats" could possibly know. They charge that the regulations were made on the foundation of bias or of politics and with utter disregard for the facts. When one flyway receives a longer season or larger bag limits than another, the Service is accused of discrimination, although these same folks are perfectly willing to accept different seasons for resident game birds in their own states.

Occasionally, the influence of the dissidents is such that bills are introduced into the Congress in an attempt to force the granting of special privileges to some state or section of the country. Without exception, such proposals are sponsored by pressure groups who seek to lessen the restraints imposed upon them by the Migratory Bird Treaty Act and the regulations. Never are they designed to give added protection to waterfowl. Although aggravating to the administrators, such legislative attempts have some measure of benefit in the long run. They quickly bring together the powerful friends of wildlife, both in the Congress and without, and in no single instance has there been a successful attempt to weaken the protective structure that was so many years in the building. Now, the proposals of selfish pressure groups merely inspire the growing army of sincere conservationists to greater unity in opposing harmful legislation.

CHAPTER IX

Wildlife "G-Men"

*T*HROUGHOUT the 48 states and Alaska the Fish and Wildlife Service maintains a corps of alert game management agents whose duty it is to apprehend individuals who violate the Federal game laws. In addition, the Federal agents cooperate with state game wardens in catching violators of state laws and, in turn, the states help the Federal agents nab their men. With the cooperation between Federal and state game agents the country has, in effect, a small army of men protecting the nation's wildlife resources.

Rounding up smuggling rings and wild duck bootleggers is only part of a Federal game agent's work. During the hunting season he must check sportsmen's guns and daily takes to see that all hunters abide by regulations.

Most of the sportsmen, however, are cooperative because they realize that these public servants are helping to keep up the bird populations by checking all comers to detect the comparatively few violators who exceed bag limits. That sportsmen do realize the necessity of having alert game agents and do appreciate the value of this work is indicated by the fact that sportsmen's organizations are constantly asking Federal game agents to speak at their meetings.

This, by the way, is another reason why the Service is exacting in the requirements for game agents. An agent not only must know a great deal about the outdoors, how to apprehend violators, and how to present his cases in court, but he must also be able to speak before sportsmen's groups, conservation clubs, school assemblies, and in other meetings.

In this branch of the Service's work we have aviators, engineers, pharmacists, newspapermen, salesmen, professional auto racers, lawyers, expert mechanics, and musicians. They were all successful in those fields, too, but the lure of the outdoors and their interest in conservation attracted them to this organization.

Despite their varied qualifications, all Federal agents have these abilities in common: They must first know wildlife; then they must be competent in handling motors, boats, and cars; must understand the use of firearms; be diplomatic; know the law, and know how and when to make arrests. Former experience is often helpful. In fact, few agents are employed in the Fish and Wildlife Service who have not had two or more years of experience in enforcing state fish and game laws. Some peculiar qualifications often assist. For instance, a few years ago a ring of bootleggers were apprehended by a game agent who before entering the Service had played in swing bands appearing in large theaters in the west and midwest. He had become interested in game management work and finally was accepted as a full-time agent. His former profession came in handy when a ring of bootleggers operated through a popular night club. The bootleggers, however, knew the regular Federal agent and hid their illegal ducks in a secret ice box whenever he was seen in the vicinity of the club.

One night the club's orchestra had a new "hot" drummer. A few days later the regular Federal agent for the district walked into the night club and went directly to the secret ice box where he found wild ducks that were used as evidence in Federal Court a few days later.

The "hot drummer" was a Federal game agent working on an undercover assignment.

Undercover work is probably the most difficult task that can be given a Federal agent. This type of activity is usually directed against rings of operators violating game laws on a large scale. The agent or agents assigned to this work appear in the vicinity of the area posing as any of a number of persons. During their stay in the vicinity the undercover agents are expected to gather evidence that will break up the illicit activities.

It isn't an easy job. Agents must work all hours of the day and night and in all kinds of weather. Once they are on a case, they can't quit until it is closed.

The Service can point with justifiable pride to the achievements of its "G-men". For example, several years ago Chicago's duck bootleggers laughed at the Federal game agents who were on their trail. For a year the smugglers had been buying and selling wildfowl and still the game agents couldn't track down the ring of violators.

The smugglers had a stooge, an old waterfront derelict who didn't know what was in the package he carried from one illegal dealer to the other. One night, four bootleggers met in a dark alley. A large package and green bills changed hands.

"Thanks, Dopey," one of the men whispered. "You know where to get the next one tomorrow night. Here's a buck for your trouble." So Dopey shuffled off with his dollar bill—a marked bill.

Two minutes later Federal agents swooped down on the cluster of duck bootleggers. The next day, when the leaders and members of the Chicago ring were in Federal Court, the principal witness was Dopey, the waterfront bum. But Dopey was sober and well-dressed, for the erstwhile down-trodden bum was John Perry, game management agent of the Fish and Wildlife Service.

During World War II the talents of the U. S. game management agents shone to particular advantage as they were called upon to assist other law-enforcement agencies in the job of protecting life and property and the safe-guarding of this country's interests. In remote coastal areas, with which Service men are so thoroughly familiar, they furnished patrol boats and acted as guides for the U. S. Coast Guard. They reported enemy aliens; found and reported short-wave radio stations being operated unlawfully in restricted coastal areas; controlled gulls at two Army airports where these birds had become a serious menace to safe flying; arrested and obtained the conviction of persons who stole Government property; acted as instructors to first-aid classes; helped extinguish fires on National forests; reported draft evaders; reported draft-age hunters who failed to have their Selective Service Registration Cards; taught seamanship, navigation, and safety at sea to men in the U. S. Coast Guard Auxiliary; apprehended a railroad bridge guard who had deserted his post of duty and was hunting ducks during the closed season; worked with the Army Intelligence Service regarding the cutting of levees during floods; detected a man who was trading liquor and narcotics to soldiers for ammunition which they had stolen from the Government; reported to the Navy Intelligence Service the theft and sale of Navy automatic rifles; assisted in the capture of escaped German prisoners of war; personally apprehended a German spy, a graduate of Heidelberg University, in the act of photographing defense plants and turned him over to the FBI; and reported the full facts concerning a German alien doing espionage work while wearing a U. S. Army Air Corps officer's uniform. In the West they even apprehended cattle rustlers.

For anyone who might have the mistaken idea that the life of a Wildlife agent is soft and easy or humdrum and dull, a perusal of the files of any good wildlife protective agency will soon prove to the contrary. There is plenty of excitement in doing the everyday job of enforcing the laws and regulations. To illustrate the point, I have taken from the files of the Fish and Wildlife Service some of the more outstanding reports of these fearless and intrepid employees. Their statements, unconcerned and matter-of-fact, reveal drama and excitement aplenty. The usual method of submitting reports to superior officers is by means of the so-called "memo." Here are a few good samples:

> Memo to Regional Director L. L. Laythe, Region No. I,
> From Game Management Agents Hugh M. Worcester and C. G. Fairchild.
>
> Berkeley, Calif.
> March 9, 1949.
>
> Subject—Market Hunting case of Don E. Smith, E. L. Ziegler, and Albert Ford.

As you have been previously advised by telephone and by wire, Service Agents working in the closest of cooperation with wardens of the California Fish and Game Department have just completed a haul that tops any case ever made in California, and we think may also hold the U. S. record. We not only bought 1,006 birds with the special funds you made available to us, but we also caught the violators red-handed.

You requested that our report in this case include some background information since, as you recalled, Don Smith had been involved in illegal waterfowl operations on previous occasions. You were correct. We first obtained a lead on him in the fall of 1946. A prominent business man in Sacramento had been purchasing wild ducks from Donald E. Smith and had arranged for the purchase of 100 wild ducks to be delivered at 10:00 a.m., January 16, 1947.

Agents Fairchild and Worcester obtained a search warrant and at 9:00 a.m. on January 16, accompanied by State Game Wardens Gene Durney and H. S. Vary, arrived at Don E. Smith's residence and searched same. In his home we found also Curtis Golden, Smith's brother-in-law, and Catherine Stanley, sister of Smith. These three persons were at that time picking wild ducks in a small kitchen in the residence, which consisted of an upper flat. These three persons were arrested and we seized 169 wild ducks and 4 wild geese. Later, Don E. Smith was found guilty in Federal Court, Sacramento, by Judge Roger T. Foley, and was sentenced to County Jail in Fairfield, California, for 90 days and was fined $750. He served his normal sentence and took a pauper's oath. In this same case Curtis Golden plead guilty and was sentenced to 30 days in the County Jail and paid a fine of $150. Catherine Stanley paid a fine of $50.

One year later, in January of 1948, an informant called at Agent Fairchild's home and stated that Donald E. Smith had a load of ducks and could be contacted at a Sacramento telephone number. With the assistance of two Deputy Agents and the Under-Sheriff, as well as a State Game Warden, we moved in at 5:00 a.m. on January 14, 1948 and arrested the same Don E. Smith and Bert G. Coffer, with 176 wild ducks in their possession. When the case was brought before Federal Judge Lemmon in Sacramento, Smith was convicted on 3 counts, fined $500 and given 6 months in the Federal Penitentiary at McNeils Island, Washington, and placed on probation for 5 years. Coffer was convicted on 3 counts, fined $300, and placed on 2 years' probation.

This, apparently, had no lasting effect on Smith's duck bootlegging activities, since on February 6, 1949, we were informed that Don E. Smith was in San Francisco, with 120 wild ducks in his car. With this tip, we soon made arrangements through the California Department for one of their reserve Wardens, Mr. John C. Pettey, to make a purchase of ducks from Smith. Telephone contacts were soon made with Smith at the residence of E. L. Ziegler in Sacramento. Soon a call came through to Mr. Pettey from a man who identified himself as "Zeke", and who advised that he could deliver 180 wild ducks at once. Given a little time, he assured that he could secure another 800 or 900. The quoted price was $2.00 per bird. Arrangements were made for delivery of the 180 birds at the Sacramento Municipal Airport. Accompanied by Federal Agent Elder of Los Angeles, Warden Pettey in an expensive plane took off for Sacramento where "Zeke" was to meet them. He was to identify himself by carrying a card in his hand. He, of course, had no idea that the plane had been loaned to us by a public-spirited sportsman of Long Beach, prominent for many years in the California Ducks Unlimited program. Neither did he know when he transferred the 180 ducks from his car to the plane and accepted $360 in cash that the bills were marked for later identification. Ziegler inquired as to whether Pettey could use another 2,000 ducks with the feathers on and was informed that any number would be welcome providing the price was right.

On Sunday, February 13, Pettey received a phone call from Ziegler, stating that on the night before he and some friends had shot and killed over 500 ducks and they should have at least 1,000 by Monday. Arrangements were made for delivery of all he could supply and the Federal and State Agents immediately began laying plans to make a real haul. A panel truck painted white and carrying an ice cream sign was rented in Berkeley. It was driven to Sacramento to be used in picking up the ducks. Arrangements were also made in Sacramento to obtain a good-looking car for Elder's and Pettey's use upon their arrival from Los Angeles.

Approximately 20 Federal and State Game Agents assembled at Agent Fairchild's office at 9:00 a.m. on Monday, February 14, dressed in civilian clothes, and were briefed as to their parts in the case. A warrant was secured for searching E. L. Ziegler's home in Sacramento. Five two-way radio-equipped State Fish and Game cars and a State airplane were used. Walkie-talkies were assigned the wardens to augment the radio cars. The exact spot where the big delivery was to be made was unknown, but we were well equipped to keep in touch with each other.

Soon, one of the wardens who had tailed Pettey and Elder to Marysville where they were discussing arrangements with Ziegler, reported that Don Smith had joined the party. Shortly after lunch, the plane which had been droning around overhead, advised that the white truck and two other cars were leaving the front of the Marysville Hotel proceeding towards Chico. Soon we were advised that a red car, presumably guarding the rear flank, was following the white truck. At one point the red car stopped to observe whether the party was being followed and we were instructed by the observer in the plane to carefully detour around the lookout. This we did in the town of Live Oak. Soon we were advised that the white truck and the cars carrying the buy-crew were turning into a large dairy ranch. Shortly we were informed by the plane that a pickup truck was leaving the ranch house, and was proceeding westward toward a small group of oak trees; that the pickup had stopped under the trees and that two men were loading the truck with heavy sacks. Soon we were informed by the plane that the truck had returned to the ranch and that the sacks were being transferred from the truck to the white panel truck.

At this point, one of the boys with a walkie-talkie shouted that someone was calling for a warden, and plenty of us swarmed in, I can assure you. Dashing up to the ranch, we deployed about the buildings in a hurry, ready for any eventuality, but we soon found that Agent Bud Elder, Pettey, and State Warden Gilbert Davis who had been driving the white truck, had already placed E. L. Ziegler, Donald E. Smith and Albert Ford under arrest and had them handcuffed. These precautions were taken because

Donald Smith had stated that he would not stand another arrest. Moreover, Ford was very nervous and it was felt that it might be dangerous if he became at all suspicious that this was not a real transaction. Pettey was also informed that one of the members of the gang had killed over 50,000 wild ducks for him, Ziegler, during the last 4 years. After the arrest, the men were searched and on Donald Smith was found a $50 bill that had been paid to Ziegler at the airport when the ducks were purchased from Ziegler on February 10.

On March 7, 1949, in Federal Court at Sacramento, Albert Ford pled guilty to an information charging six counts of buying, selling and transporting wild ducks on different occasions. His case was referred to the Probation Officer.

Informations were filed covering a total of six counts against Donald E. Smith, and 5 counts against E. L. Ziegler for possessing, selling, transporting, and killing of wild ducks on three different occasions. Smith was not represented by counsel and the Judge appointed counsel and set his case over to March 21.

E. L. Ziegler was represented by counsel, and his case was set over to March 21.

We have some interesting additional leads on others who have been buying ducks from this ring, so it looks like lots more excitement in the near future. That special under-cover money was certainly a big help.

(Signed) Hugh M. Worcester
(Signed) C. G. Fairchild

Memo to Director Albert M. Day, Washington, D. C.

Portland, Ore.
May 11, 1949.

From Regional Director Leo L. Laythe
Subject: Market hunting case of Don E. Smith, E. L. Ziegler, and Albert Ford.

The above Migratory Bird Treaty Act cases were brought to a successful termination in Federal Court at Sacramento, California, during the April, 1949, term. Donald E. Smith was found guilty on 5 counts and received a jail sentence of 30 months and a fine of $2,500; E. L. Ziegler pled guilty to 5 counts, was fined $1,500, plus a jail sentence of 12 months and was placed on probation for 5 years. Albert Ford was fined $1,800 and placed on probation for 5 years.

I was interested in your press release announcing that this is the most serious violation of the Act since the Treaty was signed in 1918. It is appalling to think of how many birds this gang has slaughtered to supply the illicit markets in San Francisco and other California towns. Much credit is due the Agents who did this job. The State Fish and Game Department and our own Federal Agents couldn't have worked any more

Plate 16. California Pilot Warden Rene De Loach, and U. S. Game Management Agent James Birch count and sack 624 of the 1,006 wild ducks seized in the California raid in February 1949. Photo by *Sacramento Bee*.

closely as one unit than they did on this case. Several prominent sportsmen also gave us much assistance. I am glad to see that you have already publicly acknowledged and praised this aid.

To a legitimate hunter who has huddled patiently in his blind all day long trying to scratch out four or five birds, slaughter such as described in this incident seems utterly fantastic. Yet California market hunters often kill literally hundreds of ducks in a single night's operations. They long ago developed a mass killing technique locally known as the "drag." Several shooters sneak onto massed ducks feeding in the rice stubble and, at a signal, all start firing in a ground-sluicing maneuver. As the survivors take wing; the hunters pour lead into them with automatic shotguns, some of which have extended magazines that hold as many as 10 shells. The kill is appalling. Obviously, such slaughter is not for sport. It exists because people are willing to pay high prices for illegal duck dinners. There will always be those willing to conduct this bloody business so long as folks insist on creating markets for wild game.

The life of a game agent is not easy, although from the number of applications received by the Federal and state law enforcement agencies each year, I am convinced that few people realize how tough the job really is. The Federal agents concentrate much of their effort on apprehending the more flagrant game violators. To be sure, the checking of hunters for state licenses and duck stamps, and to see that guns are in proper order, are all a necessary part of any good job of enforcement. Agents must also run down depredation complaints in areas where waterfowl are destroying crops. These Federal workers carry the brunt of the program of protecting agricultural areas from depredations by waterfowl.

The Federal staff has always been small and will probably continue to be so. The state departments furnish the bulk of the personnel, while the Federal agents work with them closely to correlate the waterfowl enforcement activities. With the cooperative attitude of the majority of states, the use of state wardens to assist in enforcing the Federal regulations, and the services of the Federal agents in aiding the states in the enforcement of the Lacey Act make a very wholesome and effective operation. Many Fish and Game Departments now have laws which prescribe that the state regulations for the protection of migratory birds will be the same as the Federal regulations. Thus, violations of the Federal regulations automatically become state offenses—punishable either in state or Federal courts. This wholesome, reciprocal approach has been an effective means of tying the enforcement work of the state and Federal agencies together. Although the great majority of the minor infractions of the regulations are handled in state courts, the Federal agents are highly effective in breaking up the market hunting that still persists in some portions of the country.

Breaking up these duck bootlegging rings is often dangerous business. In many areas, the local folks still feel that they have the right to kill any quantity of ducks how and when they please, just as did their forefathers. Duck trappers, poachers, and commercial killers are always armed. When an officer suddenly appears out of the marsh, almost anything may happen. Both parties are far from human habitation and, to a real outlaw, a quiet marsh threaded by dark and muddy channels sometimes becomes more tempting as a final resting place for a game warden than does the cold cell of a jail for the violator. Agents must be fearless, quick-witted, and able to take care of themselves in circumstances which are often most trying. The Federal agents are instructed to be courteous but firm in apprehending persons who have failed to observe the regulations, but when they happen to come across some hot-tempered, armed individual who does not submit to questioning or who resists arrest out on the lonely marsh, they need cool heads and stout hearts. For example:

Memo to Regional Director John Pearce, Region 5.   Nov. 28, 1948
From Leon D. Cool, Jr., Leonardtown, Md.
Subject: Patrol on Smith Island, Md.

On the evening of Nov. 26, 1948, Agent Roy R. Ferguson and I flew the Piper Cub to South Marsh Island, Md., with the intention of remaining in the area to attempt to catch a duck trapper on Smith Island. A short time before dark, we heard five shots so flew across the Island to look for the hunter. We did not find the hunter but did find a baited duck trap.

The next morning, Nov. 27, 1948, Agent Ferguson and I took off with our seaplane as soon as it was light enough to see and flew to the location where we had found the trap the evening before. We were just a few minutes too late as a boat with two men was just leaving the marsh where the trap was set. We landed but it was too rough to go alongside the boat without taking a chance of wrecking the plane.

We then took off and went to Smith Island. As we flew over some traps that had been baited the night before we saw a man leaving one of the traps with his hands full of ducks. We landed and apprehended him before he could get his motor started. He turned out to be a 14-year-old boy from Rhodes Point, Smith Island. We took him back to the trap and destroyed it and picked up the ducks, 22 in all. Due to his age, we decided to take him to town and talk with his parents. Upon arriving at Rhodes Point, we were met by a hostile mob of about 25 men. They either would not let the boy's mother come out or possibly she would not, at least we were unable to see her and were informed that his father was working elsewhere and they did not know when he would return.

We received a lot of profane, verbal abuse from all except one or two of the crowd. It was impossible to reason with them and they made numerous attempts to start a fight with us. It was impossible to know the names of the people we talked with but the crowd in general made it clear that they intended to trap ducks on Smith Island and that if we didn't stay away from there someone was going to get hurt. They also stated that at the first opportunity they were going to ram the plane with a boat or wreck it in some other way.

We tried to reason with them and some of the younger fellows were inclined to try to be friendly but the older men and ringleaders would not permit it. They thought we were going to take the 14-year-old boy to jail so they took him away and said they would like to see us get him. We informed them that if we wanted him we would send a U. S. Marshal after him.

We were in the town of Rhodes Point for about two hours and when we left it was necessary to return to South Marsh Island to get the other plane as the one we had was almost out of gas. We immediately returned to Smith Island and started to destroy duck traps. The Smith Islanders evidently expected us to return with boats and more men as they had their large boats anchored in the channels to the towns and other boats cruising around the creeks. Every time we landed to destroy a trap they tried to cut us off and had we been operating with a boat they would have succeeded. As it was, it was rather close a couple times. There also were four shots from a high power rifle fired from town at the plane while it was on the water and we were destroying traps. We destroyed 12 baited and working traps and banded and released two black ducks. We then returned to South Marsh and destroyed the trap there.

I informed the Smith Islanders that we were going to stop the duck trapping on Smith Island if we had to pitch a tent and move in.

And the destruction of duck traps continued throughout the winter, even though two months later a bullet from a high-powered rifle landed squarely between these two Federal agents while they were breaking up a duck trap. Mud splattered all over both of them. But the prices for wild duck dinners in a few restaurants in Balti-

Plate 17. Federal Agents banding ducks found in illegal trap on Chesconnessex Marsh in Virginia. These traps have long been the chief source of supply for the illicit duck racket on the Atlantic Coast.

more, Philadelphia, New York, and Washington were bound to reflect the scarcity of bootleg waterfowl as the wrecking of the duck traps went on.

The traps are usually about 5 by 4 feet in size and 18 inches high, made of poultry wire. Many trappers take an average of 15 ducks nightly, although others take as many as 40. An indication of the extent of the illegal slaughter of waterfowl by this means is the fact that more than 500 such contraptions were destroyed during the winter of 1937-1938. A similar program of trap destruction has since been continued annually, resulting in a saving of untold thousands of waterfowl and a material reduction in the number of traps.

And another example from the Eastern Shore:

Memo to Regional Director John Pearce, Region 5.

Chincoteague, Va.
Feb. 25, 1947

From John H. Buckalew, Refuge Manager, Chincoteague National Wildlife Refuge.
Subject: Assault of Refuge Personnel.
Reference is made to our telephone conversation of February 24, in

which I advised you of a felonious assault on my person by two hunters who had been violating the refuge.

Before daylight on the morning of February 22, I made an attempt to apprehend two or three persons who, according to information I had received, had been going on the refuge and killing waterfowl. At approximately 1:30 p.m., I was walking north along the road leading through the woods bordering the side of the refuge. I had stopped and looked back along this road for possibly a minute or so. When I looked up the road I saw two men starting to run across it toward Assateague Channel, on the west side of the island. Both men had shotguns, and one carried a bag containing a number of ducks. When these men found that I had seen them they started running back into the thick brush, dropping the bag of ducks. At the time I saw them they were approximately 75 yards from me. I told them to stop and started in pursuit.

Within a very short distance I had gained on them until only about 30 to 35 yards separated the hunters and myself. At this time one of the men jumped behind a tree and fired, the charge striking me in the left side and chest. As I ducked toward the shelter of another tree he fired again, the charge this time striking me in the face and arms. Before I could recover from the shock of the shots and clear my eyes of blood both men had disappeared in the thick brush. I continued my patrol in an effort to locate these men, but failed to find them until approximately 5:30 p.m. Then realizing the futility of trying to find these men, and not knowing the extent of my wounds, I returned to Chincoteague and went to the hospital at Nassawadox, Virginia, where a number of the shot were removed and I received treatment for any possible complications that might arise.

(Signed) John H. Buckalew

P. S. You may find a number of errors in this and other correspondence for today, as I have not yet regained the use of my right hand.

Fortunately, the wounds which Mr. Buckalew suffered were not fatal, although they were exceedingly painful. The Baltimore office of the Federal Bureau of Investigation was immediately contacted and soon entered the picture. An FBI agent, accompanied by a Virginia trooper soon had the outlaws in hand; the case was tried in Norfolk, Virginia, on May 16, 1947, and Walter Powell Clark, of Chincoteague, Virginia, was sentenced to two years in a Federal prison on charges of assaulting a Federal officer with a deadly weapon, with collateral charges of trespassing on the Chincoteague National Wildlife Refuge, of hunting and taking waterfowl on the refuge, and of killing ducks during the closed season.

The personnel on National Wildlife Refuges, responsible for the enforcement of the Federal laws and regulations, must be constantly on the alert to detect any illegal hunting or trespassing. That this assignment is frequently a dangerous one is evidenced by a recent incident on the Okefenokee National Wildlife Refuge, Georgia.

This refuge, embracing as it does the major part of the Okefenokee Swamp, famed in song and story as a unique wilderness, offers a real enforcement problem. In the spring of 1945, two refuge patrolmen, when investigating the continued setting of fires in the depths of the swamp, surprised a wizened, barefoot, little swampland farmer, one Oliver Thrift, poaching on the area. The violator started running when he encountered the officers and, as he afterward admitted, ran as far as he could, and then when he could run no further, he hid in the brush and shot both men with his shotgun as they came into view. Both were killed, although their buckshot-riddled bodies were not found until the following morning by the son of one of the officers.

The murderer, when finally apprehended, proved to be a local resident of the general area who had a grudge against the Refuge because he felt it harbored bear, and also he had been "feuding" with the family of one of the agents for several years because he blamed them for causing his arrest on minor charges two or three times in the past. He is being punished for his crime, having been sentenced to life imprisonment on two counts. Service personnel now patrol the swamp with sidearms ready for immediate use.

Religion and eel kegs figured in an interesting situation in 1935 in the Eastern Shore section of Virginia and Maryland. This is one of those regions where the natives have long felt that their right to hunt, kill, trap, and sell ducks should never be questioned. A certain rather prominent manufacturer in Chicago became entangled in the toils of the game laws and confessed that he had been buying ducks from Cape Charles, Virginia. He related how, enroute from Norfolk, Virginia, to Philadelphia, Pennsylvania, a short time previously, he stopped in a barber shop in Cape Charles, Virginia, and in the course of conversation made mention that he would like to purchase a few wild ducks. The barber showed no interest at the moment, but after the customer left, an individual who introduced himself as "Bill" stopped him and inquired whether he cared to purchase some wild ducks. The deal was made. "Bill" was given some money and the Chicago gentleman's business card on which was printed his name and address. Bill was requested to send the wild ducks to that address.

Plate 18. Confiscated battery, swivel and punt guns in the Fish and Wildlife Service collection of outlawed machines of destruction. The longest cannon measures 10 feet and carried 1½ pounds of BB shot. Its record was 112 ducks at one blast.

During the following month, one keg of "eels" was delivered to the Chicago address. The gentleman confessed that, when the "eel keg" was opened at home, it contained 10 wild ducks each weighing about one-and-a-half pounds fully dressed, with some of them containing shot. They were wild ducks without a doubt. He was very honest in his statements, telling the Federal agents that he had noticed considerable illegal traffic in wild ducks in that section at the time he had visited it, and that "Bill" had told him that at any time wild ducks were wanted he could furnish them in any number. Another order would have been placed but he did not know "Bill's" last name or address.

This started a chain of events which soon brought to the quiet little coastal community a young photographer, presumably connected with one of the southern educational institutions, who was interested in securing photographs of native birds for a publication which he was writing. He was finally accepted by the local folks and became acquainted with many who were working in the duck-trapping and market hunting traffic.

This young stranger was also something of an evangelist. He gained the confidence of the local minister, and together they began preaching the story of conservation and protection of the ducks and geese that always frequent that section of the country during the winter. They were so successful that they were responsible for having removed from the locality several large-bore punt guns, each 8 to 10 feet long, some of them weighing as much as 175 pounds. He was not long in learning the identity of "Bill," the same chap who, using the name of a deceased relative, had been shipping waterfowl in eel kegs to Chicago and many other parts of the country.

The young photographer turned out to be a Fish and Wildlife Service under-cover agent, one R. D. Hildebrand, whose testimony, together with that of other Federal agents, resulted in a Federal Court sentencing one William Powell to six months in jail on one count and six months on each of four other counts, to run concurrently. The sentence of 24 months was suspended for five years, to be served in the event the defendant violated the game laws during that period. The murderous battery and punt guns that were turned in voluntarily by the islanders now make up a part of the Fish and Wildlife Service collection of illegal weapons. Incidentally, Agent Hildebrand has not felt it prudent to return to the locality since that time.

The files of the Fish and Wildlife Service contain scores of interesting accounts of the trials and tribulations of game agents enforcing the Federal Migratory Waterfowl laws and regulations, but a good example of the patience, level-headedness, and cool nerve that must sometimes be displayed is well recited in portions of an affidavit submitted by Robert S. Bach to his immediate superior, Hugh M.

Worcester, at Berkeley, California, on February 15, 1936. This follows:

Deponent says:

That on February 15, 1936, I, together with Deputy Game Management Agent James Gerow was assigned to investigate reports that William J. Sexton was killing wild ducks in the vicinity of Pinole, California. That at 5:00 p.m. I observed William J. Sexton with a shotgun in his hands, standing by his house and a woman known as Vera was standing there with him. That an estimated 250 canvasback ducks were feeding near the water's edge and in the vicinity of the William J. Sexton property. That Deputy Agent James Gerow and I were concealed under a live oak tree which is located on a bluff in the vicinity of the Sexton property. That at 6:40 p.m. five shots were heard very distinctly and they came from the rear of William J. Sexton's property. That Deputy Agent Gerow and I investigated. That together we entered the property of William J. Sexton and divided, Deputy Gerow went one way, I another. That I approached the water's edge and observed William J. Sexton rowing a boat toward the shore. That I crouched down and waited. That William J. Sexton tied up his boat and shouldered a pair of oars, then started to wade ashore carrying four of the wild Canvasback ducks which were shot. That I walked up to William J. Sexton and showed him my Deputy Game Agent's badge and said to him, "I am a Federal officer and I am placing you under arrest for the possession of wild ducks, you must come along with me." That William J. Sexton resisted arrest by jabbing a pair of oars in the pit of my stomach, the blades forward, that I took one of the oars away from him and threw it away out of reach, that he again raised the other oar, this time up high and struck me a terrific blow over my left shoulder. That an oar in the hands of an enraged person is a deadly weapon.

That it was necessary that I draw my pistol in my defense, that I used the pistol as a baton and did have to strike William J. Sexton over the head which did not knock him out. That Deputy Agent Gerow approached at this time and picked up the four canvasback ducks that William J. Sexton had dropped before he had struck me with the oars. That when William J. Sexton observed Deputy Agent Gerow he again struggled with me. That it was necessary that I had to remove him bodily from the marsh. That William J. Sexton did say, "The boy ran up to the house with the gun after these ducks were shot."

That William J. Sexton was allowed to enter his house after he had promised to give us the shotgun that killed the wild ducks. That Deputy Gerow and myself entered the house with him. That on the inside of the William J. Sexton house, William J. Sexton did say, "I am in my own house now, you fellows can't do a thing, I am not going to give you the gun and I am not going with you, I want you to get out of here and leave me alone."

That William J. Sexton reached for a double barreled shotgun but was overpowered.

That William J. Sexton did seize from a table a hand axe which is a deadly weapon and did strike me several blows. That one of the blows injured my right hand when I reached out to intercept it, that several of the blows landed on my injured shoulder where the oar had hit me, swung by William J. Sexton. That in self-defense I again drew my pistol and using it as a baton I did strike William J. Sexton over the head with it. That this blow dazed William J. Sexton and I was successful in obtaining the hand axe he used in striking me. That William J. Sexton submitted to go to jail and that on leaving the room he ran into the bathroom and locked the door. That William J. Sexton defied me to come in after him and refused to come out. That it was necessary that I break the door in and again gain the custody of William J. Sexton. That another scuffle followed as he was being taken out. That William J. Sexton did have in possession a piece of iron which is a deadly weapon. That he did strike both Deputy Agent Gerow and myself with it. That Deputy Agent Gerow in self defense did strike William J. Sexton with his leather police sap. That William J. Sexton ran into the living room of his house, shouting, "I would rather die than go to jail. I know you fellows are Federal officers and once you get me in jail I never will get out." That William J. Sexton did seize a Winchester lever action .32 Cal. rifle No. 362869 which is a very deadly weapon, and did face us saying, "I will kill you if you do not leave this place." That William J. Sexton did partially raise the barrel of the rifle up and at this time I again drew my pistol in self defense and said to him, "If you raise that barrel of that gun any higher I am going to shoot, we are getting serious about this business and you had better put that gun down." That William J. Sexton did put the gun down, and said, "If you will go and get the Constable, Mr. Gene Shea, and bring him here to this house and he says you have a right to take me out of this house I will go and not cause you any more trouble."

That I went after Constable Gene Shea. That Deputy Agent Gerow stayed with the prisoner. That I returned accompanied by Constable Gene Shea of Pinole township. That Mr. Shea did say to William J. Sexton, "Billy, they had a perfect right to hit you and come in your house and you are lucky that you are alive, these men had a perfect right to kill you after the way you acted. They are Federal officers and they know what they are doing, and the best thing for you to do is to go along with them." That William J. Sexton did put on his cap and coat and did come along to the jail without any more resistance. * * * *

Following arraignment in Federal Court, this unruly gentleman was sentenced to a term of six months in jail for assaulting Federal officers and six months for killing and possessing six canvasback ducks out of season, the sentences running consecutively.

I hope the reader will not gain the impression from the records I have cited that all of the Fish and Wildlife Service difficulties center about either the Atlantic seaboard or California. That is not the case. The illegal killing of waterfowl is fairly well distributed throughout the country near large centers of population wherever birds may be illicitly sold to expensive restaurants. There are always wealthy patrons who feel that it is smart to take their friends to a night spot where all can enjoy the taste of illegal game.

Such practices are not confined to ducks alone. Quail, grouse, deer, and elk also find a ready outlet in some cities. If there were no market for game birds and animals there would be no commercial shooting. Some killing for home use by local people will probably always continue, but this is not the major problem.

A good case in point is recorded in the case history of one R. B. (Jack) Horner, who was arrested on February 23, 1938, near Fort Worth, Texas, for violating the Migratory Bird Treaty Act. The court records on file indicate that Mr. Horner started his career soon after he came to Fort Worth in 1929 as a small-time bootlegger. He expanded and built the "Gingham Inn" in 1933. The night-club catered only to the wealthier patrons and operated more or less openly in defiance of Federal and state conservation laws and state liquor and gaming laws. They further disclose that the defendant had persuaded indigent country boys to engage in the nefarious racket of supplying him with wild game of all species, had caused kitchen help to become involved in preparing the game, and had induced waitresses to serve game to his patrons.

And then, on November 9, 1937, the Federal agents and state game wardens raided the Gingham Inn and found 143 wild ducks. On the following day, Jack Horner was arrested, and the local sheriff while looking for illicit liquor found another five wild ducks cooked ready for serving. The defendant was charged with eight counts of violating the Migratory Bird Treaty Act, and was tried on November 19, 1937. All facts were admitted as alleged in the information submitted by the arresting officers, who when the trial was called, had sufficient information to add 9 co-conspirators who were indicted on five overt acts out of 160 acts that had been reported by investigators.

This case made quite a stir among the local folks who enjoyed fine game dinners, but that it was not taken too seriously is indicated by the opening argument of the attorney who defended Mr. Horner. His plea opened thusly:

Had Benjamin Franklin been able to have attained his express wish that he might be pickled in a barrel of Maderia and re-awaken a hundred or a hundred and fifty years after the drawing of the Constitution, in which he had so great and influential a part, he would doubtless be surprised beyond measure to learn that under it the central government had progressed (?) so far that five gentlemen could not sit down and express a desire for roast duck listed on a menu, to be washed down, mirabile dictu! with some wine, without a Federal officer, dressed in a tailor-made uniform and Sam Brown belt with epaulettes, embroidered thereon the legend "U. S. Biological Survey," arresting them, and they actually pleading guilty before a United States Commissioner, and being released on a promise to become witnesses against a man who had committed the dastardly crime of buying and having in his possession ducks and doves. * * * *

But the Federal officers did not feel the same. In testimony before the Court, they recommended that not only Mr. Horner, but all of the buyers of migratory birds be convicted "to show the wealthy and affluent who can buy any delicacy that the market offers that the wildlife of this country is not for sale, that a treaty is sacred, that the Migratory Bird Treaty Act is not a joke, and that 7 million sportsmen and millions of other bird lover interests are paramount to the greedy appetites of game gourmets who are too lazy to do their own hunting other than with golden shot."

The court found Jack Horner guilty and within three weeks had handed him a sentence of 13 months in Leavenworth Penitentiary for conspiracy to violate the Migratory Bird Treaty Act. Co-defendants J. D. Paige, Aubrey L. Sparks, and Jesse Porter, were all convicted as accomplices for having served as market gunners to supply the trade for Mr. Horner's establishment, but since they had cooperated with the Government, their sentences were suspended upon good behavior for three years.

Jack Horner, a short time later, applied for a pardon (which was denied), submitting in part the following statement:

"It has been a custom in my state for years to shoot a few ducks whenever they saw fit and the ducks in my possession were bought from hunters who killed said ducks. I never shot the ducks myself and never have shot wild fowl either out of season or hired anyone else to do so. I ran a dinner club near Fort Worth called the "Gingham Inn" and various hunters brought me these wild ducks which I purchased and served in my restaurant as was common and has been the practice for years in my part of the country. I believe that, to my knowledge, I am the first man who was ever convicted and sent to a Federal Penitentiary for such an offense. I recall several cases, one particular case of a man near New Orleans who was sentenced to 30 days in jail and paid a fine of $60 in Federal Court for almost identically the same offense. There have been other cases in my own and adjoining states where, at the most, low fines were imposed in recognition of the fact that such offenses were not and have never been regarded seriously by the community at large. * * * *

Catering to the peculiar appetites of wealthy gourmets who persist in encouraging the illegal take of migratory waterfowl can lead to serious trouble for the hunters who are willing to take the chance of killing protected species to satisfy a demand which is wholly unnecessary and, in fact, sometimes almost unexplainable. The case of Howard Blewitt, of California, is a good case in point. On January 28, 1936, at Los Banos, California, he was arrested for violating the Migratory Bird Treaty Act, by killing—of all things—four Sandhill cranes. He previously had been arrested for violating the Treaty Act on December 1, 1935, by selling 24 wild ducks. Tried before a jury in Federal Court at Fresno, California, on April 16, 1936, the jury found him guilty. He was sentenced to 18 months, serving his sentence in a Federal road camp in Arizona.

The story that led up to his conviction is well described in portions of the affidavit submitted by Federal Deputy Game Warden C. L. Brown, who filed the following information at Berkeley, California, on February 23, 1936:

Deponent says: That on the morning of January 29, 1936 at 7:30 a.m., in company with California State Game Warden Franklin A. Bullard, on Miller and Lux property 12 miles northeast of Los Banos, Merced County, California, I watched Howard Blewitt, age 29, of Los Banos, California, hunting wild Sand-Hill Cranes.

That Howard Blewitt at this time was using a horse as a natural "blind", crouching down behind the horse whenever in the vicinity of a flock of wild Sand-Hill cranes. That during the day and while Warden Bullard and myself watched, Howard attempted to approach several flocks of wild Sand-Hill Cranes and was at all times carrying a shot-gun.

That Warden Bullard and myself watched Howard Blewitt all day and until it was nearly dark. That at 7 p.m. Warden Bullard and myself heard Howard Blewitt shoot seven (7) times in rapid succession. That immediately after the shooting a great commotion was heard among the wild Sand-Hill Cranes.

That at 9:15 p.m. Howard Blewitt drove up to a gate in his car at a very fast rate of speed, and with the headlights turned off. That when Howard Blewitt got out of his car to open the gate Warden Bullard and I stepped up to Howard Blewitt and said, "We are Federal and California State Game Wardens, and we would like to look through your car."

That, at this time, I observed fresh blood on Howard Blewitt's sleeves. That I did seize at this time Howard Blewitt's shells and shotgun and that Warden Bullard and I proceeded to the Los Banos Refuge, north of Los Banos, California, and with Howard Blewitt under arrest. That on January 30, 1936, and at 9 a.m. in company with other wardens, I assisted in tracking Howard Blewitt's horse from where he had hidden his saddle to a point five hundred yards east and found hanging on a barbed wire fence, four (4) wild Sand-Hill Cranes. That entrails lay about on the ground beneath the Cranes. That I seized one Browning Automatic 12 gauge shotgun, 47 loaded shells, one saddle and blanket, one pair rubber boots and four (4) Sand-Hill Cranes to use as evidence in this case.

(Signed) C. L. Brown.

The Lacey Act

As previously recorded, the Lacey Act, designed to protect all wildlife against exploitation, antedated the Migratory Bird Treaty Act almost 20 years. It provides that no wild bird or animal, nor part thereof, may be shipped in interstate commerce contrary to the protective laws of the state from which taken, or into which transported. It made the powerful interstate commerce clause of the Constitution effective in so far as game is concerned and gave substantial help to enforcement in the various states. Cases prosecuted in Federal courts in recent years under the Lacey Act have covered the interstate shipment of live cardinals and buntings, deer, raw furs, mountain sheep, elk, antelope, bear, squirrels, quail, pheasants, and grouse.

One of the successful cases completed under the Lacey Act centered about one Philip Forgash, who purchased unlawful beaver pelts in the state of Idaho and then shipped them to New York, where buyers placed them in the fur trade. Fish and Wildlife Service agents working in New York in close coordination with others in Idaho and with the assistance of the state fish and game department personnel of that state, succeeded in weaving together evidence which, when presented to the court, brought on convictions that gave a severe jolt to the illegal beaver traffic in the West. Probate Judge George A. McCloud, of Haley, Idaho, assessed Philip Forgash $300 cash fine and six months in jail, and his accomplices who assisted in the purchase of the illegal skins were also given heavy punishment. James E. Reddy was fined $300 and given a six months' jail sentence, later suspended. Wade Gutches was fined $300 with 10 days in jail; Jesse Luton, $300 and 20 days in jail. In Boise, Idaho, Federal Judge Chase Q. Clark assessed an additional $650 against Gutches, Reddy, and Luton. On the other end of the transaction, Federal Judge Clarence G. Galson, of the Eastern District of New York, fined Sam Levine $1,000, and Judge William Bondy in the Southern District fined Seymour Freiman $1,250. The total cash fines paid by all of the defendants in the case amounted to $4,100, plus a total of 212 days in jail.

Some fantastic quail trapping and selling schemes throughout the South also have been broken up under the provisions of the Lacey Act. One operator who had been under investigation by Federal and state officers over a period of years ended his final transaction in the quail business with an 18 months' sentence to jail and a fine of $1,800.

Violators Take Heavy Toll

Market hunting activities—killing ducks and geese for profit to satisfy wholly unnecessary appetites—are still a major problem for the enforcement agencies of the Federal and state governments. The annual after-season toll taken by local gunners—many of them so-called "sportsmen"—represents a constant drain on the wildlife populations. In some wintering areas, it is estimated that the illegal kill of waterfowl equals the take during the open season. Shooting then is particularly disastrous, because it may well break up pairs that are preparing for the long flight northward to the nesting grounds.

Illegal waterfowl hunting hurts every single person who enjoys the sport of wildfowling. It reduces the bird populations during critical periods of the year—and it results in shorter seasons, smaller bag limits, more restrictions for everyone. If more folks would realize the seriousness of the situation, would dissuade themselves and their friends from breaking the protective laws and regulations, would go so far as to report violators to the proper officials—even appearing in court as complaining witnesses—the management of this great resource would not be nearly as difficult as it is now. A handful of enforcement agents can never do the whole job of policing two million waterfowl hunters. Those who enjoy the greatest of American sports must do more policing on their own if they want duck hunting to survive.

It seems fitting that we here record a copy of a letter written to the press by one of the over-worked and discouraged U. S. game management agents:

UNITED STATES
DEPARTMENT OF THE INTERIOR
FISH AND WILDLIFE SERVICE
BOX 790
ALEXANDRIA, LOUISIANA
FEB. 14, 1949

Mr. Max Thomas
Editor & Publisher
Crowley Daily Signal
Crowley, Louisiana
Dear Mr. Thomas:

In recent weeks our enforcement activities in the Crowley area have given rise to a welter of half truths, rumors, conjectures, etc., all of which has tended to confuse both the issue and the public. Inasmuch as inquiries are being made as to what the true situation is I felt that you might wish to know the facts.

Several months ago I began receiving complaints to the effect that in the past countless numbers of ducks and geese had been slaughtered in the vicinity of Crowley after the close of the season, that well organized groups of market hunters had operated openly and that the majority of rice farms had their creep blinds and live decoy pens in operation and that serious violations had been committed by persons normally expected to abide by the laws. The complaints stated emphatically that a similar condition would exist after the close of the '48 season and demanded that the U. S. Fish and Wildlife Service take steps to enforce the regulations.

I investigated the complaints and found that the conservationists were actually minimizing the truth. An aerial survey disclosed 35 illegal creep blind-decoy pen combinations between Crowley and Kaplan alone. In January wild ducks were being peddled openly on the streets of Crowley, night hunters were operating from dusk until dawn and an examination of blinds disclosed that most had been in use long since the close of the season.

It was obvious that a concerted enforcement effort was needed in order to combat a condition without parallel anywhere else in all of North America. Our efforts and investigations have brought about the apprehension of a considerable number of persons proving that the complaints were correct.

As you know, the Louisiana marshes comprise one of the most important wintering grounds on the continent. If the sport of wildfowling is to survive, our people must help protect the species on the wintering grounds which will in turn insure them of an increased self perpetuating bird crop in the future.

Increased shooting pressure, (400% greater than in 1935), drainage, drought, pollution, disease and market hunting all have contributed to the decreased waterfowl population. The legitimate hunter is deprived of larger bag limits and longer seasons by the activities of the violator. It seems that whenever we apprehend a game hog the sob sisters immediately contend that all the poor devil wanted was a couple of ducks to eat. A short time ago we apprehended a violator with over a hundred ducks in his possession and when he was asked why he felt he needed so many he said, "Oh I don't like to eat 'em but I do like to kill 'em." This individual prohibited the ordinary hunter from hunting on his land during the open season. We apprehended a night hunter with 6 ducks in his possession, the following morning we checked the area in which he had been hunting and found 37 cripples dead or dying of wounds, buzzard food.

One market hunter allegedly paid for his home with the proceeds of his illegal hunting while the sportsman is asked to be satisfied with 4 ducks per day.

Conditions such as exist here were corrected in most states 30 years ago with the result that all licensed hunters for the most part respect the rights of others and by doing so continue to have good hunting.

Where else in the United States is the song bird, recognized throughout the world as an invaluable benefactor to man, shot for gumbo?

Please understand that there are no personalities involved, as a public servant I am compelled by law to enforce the Federal Regulations impartially and efficiently as is humanly possible. The only reason for our existence is to help protect and preserve a tremendously valuable natural resource so that every man will have an equal opportunity to take his share.

The solution to our problem is increased conservation education augmented by impartial enforcement. Since public apathy is the worst enemy of good government the Press can be of invaluable help in the education program by printing undeniable facts proving that good conservation will pay dividends to all men.

Should you care to discuss this matter with me at any time in the future I will be most happy to meet with you at your convenience as we most assuredly need your assistance and understanding.

Very truly yours,
(Signed) CHARLES H. LAWRENCE,
U. S. Game Management Agent.

The staff of wildlife "G-Men" has nowhere near kept pace with the growing number of hunters. The result has been that in many areas increases in violations of the protective regulations have occurred. Were it not for the excellent cooperation of the large majority of the state fish and game departments, the Federal staff of 80 agents could do little to police 2,000,000 waterfowl hunters. Fortunately, the great majority of hunters realize that they must abide by the regulations if the greatest of hunting sports is to survive.

CHAPTER X

The Sanctuary Idea

*W*HY IN the name of common sense," a disgruntled duck hunter recently inquired, "does the Fish and Wildlife Service want to deliberately ruin the sport for us duck hunters who pay the freight? One would think that you folks would have more consideration for the people who finance the waterfowl program. After all, we buy state hunting licenses each year, we buy duck stamps, and many of us contribute to Ducks Unlimited. Yet what happens?" He continued, "I'll tell you what happened to me. The United States Government recently came in and bought up all the good marsh in my country, so it leaves us fellows no place to go hunting. Now the refuge pulls all of the birds into it after the first shot on opening day. It makes us pretty sore to see the birds settle down on that dratted Federal refuge with nothing for us to shoot on the outside. Personally, I am not in favor of refuges, and I know a lot of other duck hunters who feel the same."

This comment was not unusual. We hear such quite often, and I presume it is to be expected as part of our present-day mode of life. The average sportsman forgets all about ducks and geese as soon as the shotguns are oiled and put away for the winter. To be sure, he will reminisce and do a little boasting if he had a good season, or a little griping if he did not, but the waterfowl season has come and gone and there are other things to be done, so why worry? He begins to feel the urge for the old duck blind and the favorite marsh months later when the sweetgum starts to take on its chocolate hue and the first chilly breath of autumn reminds him that it is about time to touch up the decoys for another season.

Too many hunters expect to have their ducks as handy as clay pigeons. They want them ready-made. They want them to be plenti-

Plate 19. E. R. Kalmbach, of the Fish and Wildlife Service, was awarded the honor for the eighth duck stamp, with his drawing of "A Family of Ruddy Ducks." The sale that year reached 1,439,967.

ful on their particular day away from work, and then they expect more birds to be waiting for them the next time they have a chance to go out.

Somehow, ducks can't be produced with a handful of mortar like a clay bird. You can't just stack them up in handy boxes and have them there in unlimited numbers when you want them. Ducks and geese are living things. They must have food and suitable environment the year 'round—not just during the hunting season. It takes more than the purchase of a duck stamp and a hunting license and a contribution to Ducks Unlimited to provide waterfowl hunting. It requires intelligent planning, much actual management to develop and maintain habitat, good law enforcement, and current information of populations to make the program work. There are many unknown factors, and there always will be, because waterfowl production and protection must be fitted in with ever-changing patterns of land and water uses. Human pressure will always be a major unknown in this business of waterfowl management. Sanctuaries reduce the stress.

The history of refuge acquisition has been much the same everywhere. Rarely is there complete unanimity of thought and full public support for the establishment of a sanctuary in any locality. Of course, in an abstract fashion, everyone is in favor of refuges. It makes sense that ducks and geese, the same as pigs and cows, must have food and protection, so refuges are thoroughly and completely logical in the average mind. The objection comes when the particular refuge happens to include John Jones' and Tim Murphy's favorite duck hunting marsh. When that happens, John Jones and Tim Murphy are not so sure that this refuge business is a good idea! In fact, they very often decide that it is complete nonsense, and immediately become more interested in the preservation of their own private shooting grounds than they are in protecting waterfowl.

Yet after negotiations have been completed—usually voluntarily, but sometimes by condemnation—after titles have passed to the Government, after the dams and dikes and water control structures begin to function, after grazing has been reduced to the point that lush vegetation again fringes the pools and provides nesting cover for broods of downy young, after the refuge becomes a working reality, *then* John Jones and Tim Murphy become ardent supporters. *Then* they see that these areas not only provide for the ever-pressing needs of ducks and geese but also that the birds remain in the community longer because there is a place where they are welcome to rest and feed unmolested. As the waterfowl move out into the fields and sloughs of the surrounding country beyond the well-known Blue Goose markers, duck hunters soon learn that the refuge provides longer and better shooting than was possible when there was a duck blind on every bay and a bombardment all day and every day during the open season. The ducks are no longer "burned out" after the opening day of the season. In fact, successful duck clubs have for years either provided rest ponds which are never disturbed and are shot over only two or three days a week, or they provide for both rest days and sanctuaries.

These devices are designed to hold the birds on the clubs longer in order that there may be better shooting throughout the open season.

The movement to establish Federal refuges had its inception years ago, beginning in the same fashion that all other developments in wildlife protection have started—the result of much hard work and pressure by a few interested conservationists. Growing public interest, in the first instance engendered by the American Ornithologists' Union, finally culminated in the idea of setting aside an area to be used solely and specifically for birds. Federal Refuge Number One was established by President Theodore Roosevelt on March 14, 1903, covering Pelican Island in the Indian River along the east coast of Florida. It consisted of only fifteen acres, but it was the concentration point for great flocks of brown pelicans, herons, and white ibises, which were easy prey to the plume hunters of those days. The establishment of the tiny Pelican Island Sanctuary marks a milestone in conservation because it demonstrated that the United States was willing to dedicate certain sections specifically for the use of its wild citizens. Pelican Island today has little but historical significance. In fact, the pelicans completely deserted the refuge for many years, although during the past decade they have again begun to concentrate there.

That refuge was soon followed by several others along the Atlantic coast, all of them coastal nesting islands ranging from two to as much as 200 acres in size. Others of similar nature were established off the coasts of Washington and Oregon in the early 1900's for the protection of colonial birds such as cormorants, auklets, petrels, and puffins.

These were mere tokens, the forerunners of an important principle of waterfowl management. It took a crisis to awaken the public to the greater need and the present system of waterfowl refuges was conceived in the early '30's of just such an emergency. We average Americans never stop to worry about the future so long as we have plenty at hand. For years waterfowl had been constantly going downhill coincidental with the increase in drainage for agriculture, yet we paid little attention to the effects of drainage, loss of marsh lands, and intensified agriculture on the waterfowl population until the situation reached catastrophic proportions.

After World War I, goaded by high agriculture prices, the plow turned under hundreds of thousands of acres of prairie grasses in a desperate attempt to produce more and more food. Then drouth struck. Lands that should have remained in waving grasses dotted with prairie potholes—the way the Lord intended they should be—became grim and ominous spawners of terrifying dust storms. I had an opportunity in the early 30's to drive through hundreds of miles of stifling, powdery clouds clinging to the surface of the restless prairies. I can testify that there is nothing more depressing or frightening than to live in a world of choking, grimy dust.

Severe drouth and windstorms continued from the Dakotas southward across the Great Plains into Oklahoma, Kansas, New Mexico, and Texas. Unless one had actually experienced personal contact with it, it remained an interesting topic of newspaper comment but had little significance for the rest of the country. The situation was forcibly brought to the attention of folks along the Atlantic coast, when on May 11, 1934, a reddish pall fell over the entire Central Atlantic seaboard. No one had ever before experienced such a phenomenon. Meteorologists explained that this was merely a tiny fraction of the swirling dregs of the western dust bowl. Fine particles of soil that had once produced miles of waving grasses where cattle and sheep grazed and produced their young and where ducks reared their downy broods, now had been lifted into the clouds, drifted eastward, and had been unceremoniously dropped on the city of Washington and the surrounding country.

When that happened, the dust bowl became a symbol that led to national concern for the welfare of the soil. Out of that awesome red cloud that settled on the Capital was born the Soil Conservation Service with authority and funds to try to do something to check the ravages of wind and drouth. Out of it also came the public determination to do something for waterfowl that were in desperate plight because that same withering hand of drouth had eliminated the sloughs and ponds and potholes that had given life to the broods of ducks and geese and shorebirds that at one time had been produced by the millions in the prairie states.

Public interest in the plight of waterfowl became so great that President Franklin D. Roosevelt, early in the summer of 1934, appointed a special civilian committee to investigate the whole situation and to advise the Biological Survey and the Secretary of Agriculture of means that should be adopted to relieve the desperate situation. J. N. Darling, the famous cartoonist and then Chairman of the Iowa State Fish and Game Commission; Thomas Beck, influential publisher; and the late Aldo Leopold, outstanding educator and research worker in the field of wildlife management, constituted the panel of diagnosticians. This committee promptly called for all of the data in the files of the Biological Survey which might indicate areas that could be restored and again devoted to the needs of waterfowl. Fortunately, the Bureau had been collecting such information for the previous 20 years, and had examined thousands of areas throughout the United States looking toward the day when public interest would be aroused to the point of actually doing something to make restoration possible. This special committee also requested all of the state fish and game departments, as well as interested sportsmen throughout the country, to send in suggestions. The data were analyzed with care and out of it came a waterfowl restoration program that staggered the imagination of the conservation world. Those bold dreamers had the temerity to suggest that as much as *25 million dollars* be spent in acquiring and reflooding once productive marshes.

"Ding" Darling did such a splendid job on the committee that he was requested to take charge of the Biological Survey as its Chief, to replace Paul G. Redington, who was in poor health at that time and who left the Government service on disability retirement shortly thereafter. Darling's enthusiasm, his ability to paint word pictures as dramatically as he had always drawn cartoons, his national prominence, and his drive, did something that probably no other man in the United States could have done. With the staunch support of wildlife's friends in the Congress, including such great conservationists as Senators Harry Hawes, Key Pittman, Frederic C. Walcott, Peter Norbeck, then Congressman Willis Robertson, and many others, Darling managed to secure appropriations to initiate the first constructive waterfowl program in the history of conservation.

Unheard-of appropriations were earmarked for the cause. $8,500,000 of emergency funds became available for buying lands and improving them by constructing dikes, dams, and other water control structures, for building headquarters, for fencing, and for other necessary improvements. An additional $1,000,000 was set aside by

THE SANCTUARY IDEA 151

President Roosevelt as a special fund for the purchase of refuge lands. Another $1,500,000 fund was allocated from submarginal land retirement funds for waterfowl purposes, and $3,500,000 from drouth relief funds for the purchase and development of lands within the great dust area. Another $2,500,000 was allotted by the Works Prog-

Plate 20. "Baldpates"—a pair of widgeons, with a second male coming in to join them, was the feature of the 1942 migratory waterfowl stamp, drawn by A. Lassell Ripley. Sale 1,383,629.

ress Administration to be used for engineering surveys, for water control structures, and for the development of the refuges through food and cover plantings.

Not yet satisfied with these temporary allocations of funds to do the emergency job, conservationists immediately filled ranks behind "Ding" and the other leaders and successfully fought for the enactment of the Duck Stamp Law which was approved on March 16, 1934. It provides that the receipts from the one-dollar stamp be set aside in a special fund to be immediately and permanently available for the acquisition and improvement of inviolate sanctuaries for waterfowl, for law enforcement, and for study and research into problems of waterfowl management.

As an employee of the Biological Survey stationed in Washington during the time of "Ding" Darling's two-year regime as Chief of the

Bureau, I can testify to his dynamic leadership and to the feverish scurrying of the entire staff as he drove for greater and ever greater speed in getting the program under way. His days were never long enough. I have seen him, after working at top speed for hours, sit in his office in his shirt sleeves on hot, sultry Washington nights catching up with over-due cartoons for his syndicate contracts. The strain was too great, and he was soon advised by his doctors that two years of such effort was all any human body should be expected to endure. His successor was Dr. Ira N. Gabrielson, "Gabe" to everyone who knows him, admirably trained, an able leader, and with twenty years' field experience with the Biological Survey.

When Darling came in as Chief of the Biological Survey, he brought with him a bushy, red-headed young dynamo in the person of J. Clark Salyer II, who had been doing special biological work in the state of Iowa while Mr. Darling was a member of the Game Commission there. Much of the imagination needed to initiate a national program of this sort can be credited to Mr. Salyer. He had excellent basic training, a quick imaginative mind, and a driving force similar to that of his Chief.

Also during this period, Burnie Maurek, who had been Commissioner of the Game and Fish Department of the State of North Dakota, and M. O. Steen, assistant to Mr. Maurek, joined the staff of the Biological Survey. To these two men go much of the credit for the marvelous program of waterfowl restoration in North Dakota.

Numerous others came in from the outside, while Rudolph Dieffenbach, F. C. Lincoln, Stanley P. Young, Dr. Clarence Cottam, W. Roy Dillon, and many others within the Biological Survey strained every energy to see that the program of restoration was carried forward at top speed. The result of this effort was the initiation of a system of waterfowl refuges which would eventually be established throughout the entire United States, spaced strategically along the four principal flyways. There are still many gaps that must be bridged before the waterfowl resource can be assured any great degree of permanency, but a fine start has been made. In the early stages of the refuge program, emphasis was placed upon the breeding areas because those then were the principal need. In later years, greater emphasis has been placed upon the intermediate feeding and resting areas and on the wintering refuges in the south.

CHAPTER XI

The Refuge System

*T*HERE are 282 wildlife refuges in the continental United States, of which 196 are for migratory waterfowl. The waterfowl refuges encompass close to 3,250,000 acres. The total acreage set aside for wildlife refuges in the United States and Territories amounts to about 18,500,000 acres, some 8,000,000 acres of which are included in twenty wildlife refuges in the Territories—principally Alaska. People have grown so accustomed to figures running into many ciphers when talking about Government finances or Government activities, that 18 million acres, or even three and a quarter million acres, are difficult to visualize. Perhaps a more simple approach would be of interest. There are on the Federal refuges approximately 1,600 miles of fencing, equal to a distance greater than that from New York to Havana, Cuba; there are more than 1,000 miles of roads and trails, or a stretch equal to that from Washington, D. C. to St. Paul, Minnesota. There are 450 miles of ditches for managing the water on the refuges, equal to the distance between Boston and Cleveland, Ohio.

These refuges are doing much to preserve, protect, and provide for the flights of waterfowl between the breeding grounds of the North and the wintering grounds of the Gulf coast, Mexico, and Central America. It is estimated that approximately one-fifth of the

Plate 21. Four hundred and fifty miles of dikes control water levels on the Federal Migratory Waterfowl Refuges.

continental waterfowl population uses the present refuge system during some part of the year. The program is well on its way, but far from completed. It should be at least doubled in the years ahead.

BREEDING REFUGES

Upper and Lower Souris and Des Lacs Refuges in North Dakota

Distributed along the northern tier of states, there have been established numerous Federal waterfowl refuges which provide a goodly portion of the waterfowl production within the borders of the United States. The combination of three refuges on the Souris (or Mouse) River drainage in North Dakota are really top-notch. This stream arises in the prairies of Canada, winds southward through northern North Dakota, turns to the east, and then again swings back northward into Canada where it empties into Lake Winnipeg. Its waters eventually find their way to Hudson Bay and the Atlantic Ocean. Flowing into the Souris River from the westward as it swings down out of Canada is the Des Lacs River. Both streams are in the heart of a vast glaciated expanse containing literally thousands of shallow sloughs and potholes interspersed among waving grasslands and wheatfields. This is one of the very finest waterfowl-producing regions on the continent when the melting snows and spring rains keep the potholes filled. On the other hand, when the ponds disappear in late spring or early summer they may become death traps for literally thousands of fledglings too young to fly, and of adults in helpless condition during the moulting stage.

The Des Lacs and Souris country had been famous for years as one of the great hunting sections in the United States. Early market hunters have left records of canvasback, scaup, mallards, and pintails killed literally by the wagon-loads in pass-shooting as the birds moved about from one pothole or prairie slough to another. During the days of severe drouth in the early 30's, this country lost its life-giving water, and the prairies, once alive with quacking and chattering ducks, turned into barren dust beds. "Ding" Darling, as a duck hunter, had known this section for many, many years, and when funds became available, the restoration of the Souris was one of his first projects. The net result was the establishment of a refuge system assured of permanency by a dam on the Souris that impounds 112,000 acre feet of water in a lake since named "Lake Darling," and a series of dikes which control shallow water pools in the areas below.

The Des Lacs Refuge on the Des Lacs River consists of 19,500 acres, while the Upper Souris contains 32,100 acres, and the Lower Souris 58,500 acres. All tie together into a management unit that controls the waters of the Souris and Des Lacs Rivers and holds them in reserve to be released to the nesting marshes as needed throughout the season.

After the storage reservoir had been filled and the marshes restored, the waterfowl responded in a fashion that exceeded even the fondest hopes. Those refuges have been producing birds continuously —in wet years and in drouth almost from the day the project was completed. I have had an opportunity to make some personal observations in the vicinity of the Souris Refuge system and to note its effectiveness. In the fall of 1944, Oscar Johnson, then Regional Director, and I flew the entire length of the Souris and Des Lacs Refuges and criss-crossed the intervening miles that were dotted with water-filled potholes. We took the same flight in the fall of 1945, which was the beginning of a severe drouth period throughout the Dakotas and the prairie provinces of Canada. The change was startling. Where in 1944 every pothole in the country surrounding the refuges had been filled with water and every acre of pond was liberally sprinkled with locally-reared ducks, the 1945 airplane trip showed those same sloughs and ponds as dry as the proverbial bone.

Our 1945 flight started at Bottineau, North Dakota, and again covered the three refuges. We flew the length of the impoundments on the Lower Souris Refuge, and from the air could clearly see the outline of the old drainage ditch which in 1910 had ruined one of the finest natural marshes in the country. The old canal ran straight along the Souris Valley with occasional wide symmetrical bends to conform to the river's course. That ditch had once hurried the life-giving waters out of the Souris basin of North Dakota as rapidly as was possible, but now—thanks to the Refuge—the old ditches are mere skeletons. On either side, as we flew along, we looked down on clear, shining, shallow water dotted with beds of pondweeds and emergent vegetation, and literally teeming with mallards, blue-wings, shovelers, widgeons, gadwalls, and sprig.

The six dikes across the Souris valley are spaced just far enough

Plate 22a. Tules, cattails, pondweeds and nesting ducks now replace— (see below)

Plate 22b. Swirling, shifting dust beds—and pitiful little clumps of weeds struggling for a foothold.

apart, for a stretch of 32 miles, to hold the water at the proper level for maximum food production. The pools vary in width from one-half to two miles. The banks of the old drainage ditches and the artificial nesting islands constructed before the waters were impounded were covered that day with fat birds preening in the sun. Pelicans soared below us, and an occasional flock of geese would take to the air as we approached. We crossed the prairies to the Upper Souris and then turned south to inspect the length of the unit where Lake Darling holds the waters of the river for release as needed in the marshes below. Incidentally, this impoundment has given the town of Minot the only effective protection against floods that it has ever had as well as supplying much-needed water in time of drouth.

Just below the dam are several hundred acres of fine marsh. One small section of 15 or 20 acres was particularly noticeable. Here was a fine bed of longleaf pondweed, and puddling and dipping into it was the greatest mass of ducks it has ever been my privilege to see. Years ago in Wyoming I once rode up to a swarm of bees that had decided to settle on a three-foot sage brush. Each one was crawling and pushing and crowding to retain a foothold on the precarious perch. Those ducks below Lake Darling reminded me of the bees.

The Des Lacs Refuge, created by shallow impoundments on the Des Lacs River before it empties into the Souris several miles below Lake Darling, also had a large number of birds on it as we flew over. Here again the artificial nesting islands were covered with ducks basking lazily in the afternoon sun, and the open water shallows were liberally sprinkled with birds either resting quietly or tipped down with their bills in the mud and their tails pointing straight up at us.

It was as we cut across country on the return to Bottineau that the change in water conditions from the previous year became most apparent. Here we saw numerous round splotches of dead marsh vegetation and baked mud standing out in bold relief against the wheat and barley stubble, like brown polka dots on a straw-colored handkerchief. There was no mistaking the pattern woven by the grain drills that months earlier during the planting season had skirted the edges of water-filled potholes that now were dry and mud-caked in the center and weed-bordered at the edges. Fortunately, levels had held up long enough so that in most cases the birds had passed the moult and flightless stage before the water disappeared.

Where did the ducks go when this happened? To the refuges, where man had with foresight provided for just such an emergency. This was the reason why so many birds were on the refuge pools—there was no other place to go. That was generally true throughout the entire prairie area. The only water to be found was on the Fish and Wildlife Service refuges—the large ones such as the Souris, Des Lacs, Long Lake, Arrowwood, and Lake Alice, and the 52 smaller easement refuges. If anyone doubted the value of refuges, an airplane trip over the breeding grounds in the autumn of 1945 would have readily changed his mind.

Malheur National Wildlife Refuge

Another outstanding waterfowl-producing area is the Malheur National Wildlife Refuge located near Burns, Oregon. It consists of 174,161 acres of land, a part of which was once the famous old "P" Ranch that stretched for miles along the Blitzen Valley. The Blitzen River rises in the Steens Mountains, and in those days furnished the water supply for meadows fed by irrigation ditches, with the waste waters being dumped into Malheur Lake. In its original state, and during the early stages of the "P" Ranch operations, Malheur was one of the biological marvels of the continent. It produced literally thousands of ducks, geese, and other birds, and was a mecca for western ornithologists and sportsmen. It was so famous that the area around the lake itself, consisting of some 60,000 acres, was one of the earliest refuges withdrawn by Executive Order. This occurred on August 8, 1908, and conservationists felt that a real victory had been won in preserving this magnificent area for wildlife.

It remained so until 1926. Then, due to increasing diversion of water for irrigation, with attendant evaporation losses, the lake shore gradually receded foot-by-foot and mile-by-mile. Finally the once famous Malheur Lake withered and died. It became an alkali dust bed. In the early 30's, a car could be driven over the exposed lake bottom. The nesting colonial birds and waterfowl disappeared. The mere withdrawal by Executive Order had proved to be hopelessly useless because that action had not included the control of the water supply. This marsh, once one of the finest on the American continent, became an eye-sore of glistening, stinking alkali, unfit for any use other than low-grade grazing.

Again, as has been the history with every successful conservation project, a few dreamers and idealists spent unstinted hours, money, and energy in the attempt to restore the once great Malheur to the waterfowl paradise that Nature intended it to be. Dr. Ira N. Gabrielson, who succeeded "Ding" Darling as Chief of the Biological Survey, was one of the stalwarts who led the battle for the restoration of Malheur. George Benson, William Finley, and Stanley Jewett, the latter another seasoned employee of the Bureau, were among the small band of conservationists who never relinquished hope.

When the emergency funds became available early in the Darling regime, Malheur held a high place on the list of areas to be restored. The "P" Ranch, consisting of 64,716 acres, and holding title to most of the water rights on the Blitzen River, was purchased outright. Now the waters that had once given life to Malheur Lake could again be returned to it, and the upper meadows and side streams could be developed for nesting cover and sanctuary for breeding waterfowl. The later acquisition of the "Double O" Ranch below Malheur added another fine segment to the refuge since it contains large springs with non-freezing ponds which provide open water throughout the year. Trumpeter swans from the Red Rock Lakes Refuge in Montana have recently been introduced on the pools to reestablish these fine birds on the West coast.

The restoration of Malheur Lake has proceeded even though land title for some portions of the lake is still in litigation. The waters of the Blitzen River are now keeping the upper reaches of the valley in ideal condition for nesting birds. The drainage into the lake has restored it almost to its original level. Pondweeds and rushes have returned, and so have the myriad waterfowl. As many as 1,800 pairs of Canada geese now nest there each year. The greatest colony of breeding sandhill cranes known in the United States now frequent its marshes, and twelve species of ducks nest there regularly. This area produces more diversified forms of wildlife than any other in the Federal refuge system. More than 210 species of birds and 49 species of mammals have been found on it. Antelope, beaver, mule deer, and sage grouse are found in abundance.

Although the wild things returned with unbelievable speed and in astonishing numbers, the economy of the country has been little upset by restoring Malheur to its original state. Cattle graze upon its many meadows. Haying is carried on by the local ranchers, and the refuge now feeds and pastures almost as many livestock as it did in the days when the famous "P" Ranch was operating on its own.

Yet the battle is never completely won. Conservationists must never relax vigilance if they want to preserve anything for waterfowl. Indicative of that fact was a move started in the summer of 1947 to break up the great Malheur refuge to make available some 25,000 acres of the Blitzen Valley for homesteading. This plan would take the heart out of the entire refuge water control program and could again lead to replacing the verdant marshes with clouds of swirling alkali dust. Even though the "P" Ranch management in the early days encouraged settlement and offered all sorts of favorable inducements to people who would cultivate the lands in the upper valley, farming activities proved wholly impracticable. The elevation is more than 7,000 feet. The water in the Blitzen is icy and clear as it comes from the snow-banks of the Steens Mountains to the south, too cold to encourage the quick growth needed in the short summers of this high country. Yet the urge to make things over, to constantly sacrifice excellent wildlife habitat for doubtful agricultural enterprises, is still with us. With us also, fortunately, is still the small band of conservationists who fought so tenaciously for the restoration of Malheur. They, with scores of new recruits, will not stand idly by and watch it desecrated again.

California Waterfowl Refuges

Today, if Andy Burnett could ride down out of the Sierras into the Sacramento Valley of California, he would not find those millions of waterfowl that raised to the roar of the Long Rifle as they did in 1832. Today—instead of that great expanse of ponds, each edged with tall, waving grasses and covered with thousands of wiggling and swarming ducks and geese—he would see mile upon mile of orchards and truck farms and endless rice fields. He would find the marshes gone and in their stead prosperous farms and cities dotting the lands that once belonged to waterfowl. Even in the Delta country east of the Bay he would find huge dikes and pumping plants that now suck the waters from the fields and dump them into canals which lead to San Francisco Bay. He would find the great Tulare Basin dry as a bone. Andy Burnett, in fact, would never recognize the California of today after seeing it near the turn of the 19th Century.

Plate 23a. Area before starting the Development of the Sacramento Refuge in California.

Plate 23b. Area during the Development of the Sacramento Refuge.

Plate 23c. After the development of the Sacramento Refuge.

California's problem is complicated by the fact that the breeding grounds for the Pacific Flyway waterfowl have been less disturbed by man than in any of the other great breeding areas. California's birds come largely from Alaska, British Columbia, Alberta, western Saskatchewan, and from nesting areas in the states west of the Rockies. Man has done little so far to destroy the vast marshes of Alaska. The birds there still rear their young with little loss except from storms and natural enemies such as the jaegers and other predators.

Yet a large proportion of the once productive wintering range has disappeared. As the ducks and geese come southward through western Washington, they find few places where they are welcome. The Skagit Flats country, once a magnificent feeding ground for the migrating hordes as they moved southward, has been drained and converted into one of the finest vegetable-seed producing areas in the United States. With no place for the birds to go except to the adjacent bays or into the vegetable fields, the result is a serious depredation problem. In Oregon, the stopping places have also largely disappeared, although the State Fish and Game Department has recently initiated the development of Summer Lake, and is moving forward with the restoration of famous Sauvies Island. Down through the Tule Lake and Lower Klamath country the birds find food and sanctuary on Federal refuges superimposed on reclamation projects. The waste water from the irrigated valleys in these sections go into large sumps which are ideal for waterfowl except for occasional outbreaks of deadly botulism. Feed is provided by share-croppers farming the refuge lands under permit. After leaving these two areas, the next stop is northern California in the heart of the rice growing belt.

The most important Federal refuge in California is the Sacramento National Wildlife Refuge consisting of less than 11,000 acres and including the old Spaulding Ranch where there was once one of the most famous duck hunting clubs in California. In addition to the large quantities of sago pondweed, water grass, and other natural aquatics crowding the natural ponds and sloughs on the Sacramento Refuge, hundreds of acres of rice, barley, and other agricultural crops are now grown to provide feed for the birds.

This is a most interesting and productive refuge. The pheasants run hither and yon through the tall grass as one·drives along the roads on the crests of the impoundments. Geese concentrate by the hundreds of thousands. When I first visited this area with Dr. Gabrielson during the winter of 1937, I actually could not believe my eyes. As we stood at the refuge headquarters about noon on a sunny January day, an airplane droning up the valley from the south set the geese astir. They arose from the pools and the fields on the refuge until they reminded me of swarms of blackbirds coming to their roosting grounds in the evening. The refuge manager told us that there were probably a quarter of a million geese on the refuge that day and I certainly did not attempt to dispute him.

Until very recent years the Sacramento Refuge was the only Federal area in the Central Valley, the next area being in the Salton Sea section of southern California. As rice fields continued to increase in numbers and size, as more and more areas were taken away from the ducks and geese, depredations of agricultural crops naturally became more severe. Somehow, fields filled with rice, a duck's favorite food, and with water held a few inches deep by the checks of the irrigation systems provide exactly what the birds want. The result was inevitable—a head-on collision between waterfowl and the agricultural interests. As the price of rice increased during the early stages of World War II, depredations soared to as much as one million dollars per year.

This crisis brought on a completely new approach in waterfowl management in California. A program of acquiring refuges within the heart of the rice-growing belt was initiated. Purchase was started on one unit in the Sutter Basin and on another in the Colusa Trough. The eventual purchases will amount to around 5,000 acres in each unit, matched by areas of equal size to be acquired in the agricultural areas by the California Division of Fish and Game. A program of intensive crop production was started on these lands so that the birds might be kept off of the privately-owned fields and, instead, use the feed provided for them by the state and Federal agencies.

"How", you may well ask, "does a duck know when to eat rice produced for him instead of going over into the next field to feed on grain belonging to John Doe?" The answer lies in effective "herding". Birds can be removed from fields by flares and other frightening devices, by shooting into the air above them, and by bombing. If there is some place for them to go after they have been chased from the privately-owned fields, they can be herded or driven by the use of light, maneuverable airplanes. It is fascinating to watch a good pilot swoop down on a flock of birds feeding in a rice field, and if they do not flush as a result of the noise of the plane, circle back over them and drop a small and noisy but harmless hand grenade in their midst. They come out without further argument and take to the air with loud squawking and beating of wings. The pilot circles behind them, and they start moving in the direction he wants them to go. Maneuvering back and forth in wide sweeps, he manages to keep the fleeing birds ahead of him and guides them toward the refuge. As he pulls away they settle down and soon learn that they are welcome in this particular spot even if they are not wanted elsewhere. The result of this feeding and herding program has been almost phenomenal. Damage to the rice fields has been reduced materially, and when a sufficient number of such refuges are finally established, crop damage can probably be quite successfully prevented.

This program has also been extended to the Imperial Valley where the State of California and the Fish and Wildlife Service have leased areas from the Imperial Irrigation District on which feed is grown or scattered to induce the birds to stay on the refuges. The same herding technique has been successfully used to teach the birds that they are unwelcome in the lettuce, spinach, and other vegetable fields, but that food and protection has been provided for them on the low fields bordering the Salton Sea.

Even so, the California program is wholly inadequate to meet present waterfowl needs, and the situation is growing progressively worse as more and more lands are being converted to agriculture every year and less and less water is available for ducks and geese. Privately-owned clubs have a rather large acreage in California which contributes to the environment for the birds during part of the year. The nation-wide ban on the distribution of bait as a means of luring waterfowl to the blinds, adopted during the duck depression of the 30's, put a stop to the common practice of "feeding" on these clubs. As is perfectly human, few clubs ever flooded their holdings until shortly before the hunting season and seldom provided feed after the season closed, so little in the way of food was provided except during the open shooting periods. As a substitute, clubs are now increasingly following the legal practice of planting natural foods which provide considerable in the way of habitat. The benefits of this plan obviously extend beyond the hunting season.

California will always be a problem area. Shrinking habitat because of the constant expansion of industry and agriculture is only a portion of the difficulty. The philosophy and psychology of too many of the duck hunters in California make a permanent and lasting program difficult. They have long been accustomed to the large wintering flocks of ducks and geese arriving in September and October each fall remaining through February and into March within the valleys of their state. They naturally develop a conviction of abundance. As the marshes disappear, as fewer and fewer places are available for waterfowl, the birds concentrate in the spots that are left. In fact, they locally increase even though the total flyway population has greatly declined throughout the years. Memories are short, and in the public meetings it has been surprising to hear the comments of some of the older California duck hunters who have convinced themselves, and oftentimes others of the younger crowd, that there are just as many ducks and geese in California now as there were twenty years ago. That can be true for the individual areas where there is a combination of water and feed, but those spots are now so few and far between that the over-all picture in California is really discouraging.

I flew over the entire northern Sacramento Valley with Regional Director Leo Laythe and waterfowl biologist Everett Horn in the early fall of 1947. The situation from the air was startling. We could see mile after mile of rice fields pointed up by the irrigation ditches and rice checks which wove intricate patterns on the landscape below. But those were privately-owned rice fields. Where was a duck to go if he wanted to get an honest bite without encroaching upon someone's $4.00 per hundred rice? Not many places! In more than three hours of flying over this whole section, we saw mighty little for the birds. There were two tiny spots of green vegetation on the Colusa and Sutter Federal Refuges, a larger patch as we flew over the Sacramento Refuge, and another small one to the eastward on the state's Grey Lodge Refuge. Interspersed throughout the general area we spotted a few duck clubs where water had been spread in anticipation of the forthcoming hunting season. These represented the sum total of all the places available for birds in that vast level stretch of productive farm land. Everywhere else ducks were about as welcome as grasshoppers.

Fortunately, the 80th session of the Congress enacted Public Law No. 534, pressed to passage after four years of persistent effort by Congressman Clarence F. Lea. It authorizes the Federal Government to acquire lands for waterfowl management purposes in California, to be matched with equal acreages to be purchased by the State of California. Under this Act the Federal areas may be opened to shooting by the Secretary of the Interior, but their primary purpose is to care for the birds more adequately by providing additional acres on which feed may be grown. The same Congress appropriated $250,000 to initiate the program. Ironically, the agricultural organizations gave wholehearted support to the measure while duck hunter and outdoor writer interest ranged from lukewarm support to open opposition. At the same time that the Lea Bill was enacted the State of California also made liberal funds available to the State Fish and Game Commission to acquire and develop wildlife areas. A coordinated state-Federal program is now proceeding in a satisfactory manner.

California conservationists had better begin soon to think about furnishing the basic requirements—land, water, and feed—that waterfowl must have if they are to be perpetuated. It is not enough to satisfy the conscience by a contribution to Ducks Unlimited for work in Canada. In the Pacific Flyway the wintering problem is much more acute than is the situation on the breeding grounds. We have emphasized in the public meetings, and I repeat, arguments over the annual regulations, worries about whether the bag limit should be 4, 5, 7 or 10 ducks per day, and the season 30, 40, 60 or 80 days, actually are of little moment in comparison with the more basic problems of habitat requirement. The annual hunting regulations are temporary. They are essential to sound management, but unless the duck hunters begin thinking about the needs of the birds instead of improved means of killing the largest numbers in the easiest fashion and the shortest time possible, the sport of wildfowling will some day be a thing of the past.

Mattamuskeet and Adjacent National Wildlife Refuges

In the heart of the flat, pine-clad coastal area of eastern North Carolina lies Mattamuskeet Lake. It has been the winter home of black ducks, mallards, pintails, Canada geese, and swans for untold centuries. Some 50 miles north and east of Mattamuskeet Lake is Pea Island, slightly offshore in the Atlantic. It has always been the focal point for winter concentrations of snow geese, brant, and many of the diving ducks. Twenty miles southwest of the lake is Swanquarter Bay, rich in submerged aquatics and a natural haven for scaup and other divers. All three of these areas are now Federal wildlife refuges and make a wonderful combination of units to help care for the wintering needs of the diversified waterfowl of the Atlantic Flyway. Pea Island consists of almost 6,000 acres, Mattamuskeet of about 50,000, and Swanquarter Bay of 15,500 acres.

Mattamuskeet is another refuge that has had a long history of exploitation by promotors who attempted to convert its shallow lake bed into productive farms. Convinced of the soundness of the American myth that there must be rich soil beneath any shallow lake, drainage engineers could not rest until the waters of Mattamuskeet were pumped into nearby Albemarle Sound and the lake bottom exposed to the sun and air.

The lake proper, covering about 35,000 acres, is a large kidney-shaped sump with its surface barely above sea level. Funds were raised around 1910 to launch a huge agricultural drainage project, and seven canals were cut across the lake extending from north to south. These were then connected along the southern border of the lake by two great ditches that came together in the middle of the south shore to connect with a still larger canal extending seven miles southward to Albemarle Sound. A great pumping plant equipped with four monstrous pumps, each capable of removing one million gallons of water per minute, was installed. These pumps sucked the water down the laterals into the two main canals, lifted it and dumped it into the outlet which carried it to the sea. The lake was actually pumped dry for one or two seasons, but even those gigantic pumps could not successfully battle the normal rainfall which drained into the lake from the vast flat-lands to the north and west. The surface drainage is augmented by seepage because the floor of the lake is actually below sea level. The New Holland project, as it was known, proved an utter failure, although many millions of dollars were spent in the undertaking. Several promotors fell within the clutches of the Blue Sky law, and at least one spent considerable time cogitating the hazards of agricultural drainage while resting morosely in a jail cell.

When the accelerated waterfowl refuge program took shape in the early 30's, Lake Mattamuskeet was one of the first projects to be acquired. The buildings which were by then beginning to rot and fall apart on their foundations were moved to higher ground and repaired for residences and work shops. The enormous pumps were lifted from the great brick pump house, and the building was converted into a beautiful lodge. It now accommodates thousands of visitors who come to the area to enjoy the thrill of masses of birds wheeling overhead each winter.

Others enjoy bass fishing here that is hard to beat in the East. Lake Mattamuskeet, although a Federal refuge, also provides excellent waterfowl hunting. Some 10,000 acres have always been set aside as a public shooting ground. Hunters find excellent accommodations at the lodge, while guides, blinds, and boats are furnished by the North Carolina Wildlife Resources Commission which manages the hunting on the refuge.

Feed is grown on the Federal lands surrounding the lake, and it is heart-warming to visit this spot in the winter and listen to the honking of 50,000 Canada geese mixed with the plaintive musical calls of the 20,000 magnificent whistling swans that trade to and fro in small groups above the marsh and open water. There is constant chattering and calling as the thousands of ducks and other waterfowl ply between Mattamuskeet, Swanquarter, and Pea Island. This combination of areas insures that there will always again be a spot for the wintering flocks whose ancestors had been accustomed to patronize this section. Significantly, wildfowl shooting in the whole of eastern North Carolina has also been vastly improved by this development.

Wildlife Refuges Protect Endangered Species

No less than six species of birds, the great auk, the Labrador duck, passenger pigeon, Carolina parakeet, heath hen and Eskimo curlew, are gone forever. For these the National Wildlife Refuge program came too late. All except the parakeet were hunted to extinction and it probably owes its passing to its popularity as a cage bird. The passenger pigeon is believed to have been the most abundant bird in the world, yet it was wiped out during a period of less than 50 years largely by intensive, unrestricted hunting and market shooting. If the refuge concept had developed earlier, it is entirely possible that at least some of these species might still be in existence.

The Aransas National Wildlife Refuge on the southeastern Texas coast provides the chief hope of preserving the whooping crane. Only 31 birds were found in January, 1948, along the entire

Plate 24. Trumpeter swan and duck neighbors on Red Rock Lakes Refuge in Montana.

Gulf coast, and most of these were on or near that refuge. Even this meager figure was encouraging because it was the largest number reported during the ten winter seasons since the 47,000-acre Aransas Refuge was acquired. This huge white crane may well be listed as an outstanding example of native perseverance. It faces many risks during its 3,000-mile trip twice a year to and from its unknown nesting ground supposedly somewhere north of the Great Slave Lake near the Arctic Circle. Although it has enjoyed very limited nesting success, the winter of 1947-1948 saw six brown-masked young cranes in the family groups on the Aransas Refuge. No wintering losses are known to have occurred during the ten seasons that these birds have been protected there.

The huge trumpeter swan, the fourth rarest bird now remaining in America, receives full protection on the Red Rock Lakes Migratory Waterfowl Refuge in southwestern Montana. In 1935, when this 32,000-acre refuge was established primarily for the preservation of the trumpeters, there were only 73 birds on the Red Rock Lakes areas and immediate vicinity, which now constitutes this rare bird's principal range. Through development of the refuge and management designed specifically for their protection and increase, trumpeters had been restored by 1948 to 418 in numbers. Also, 27 birds had been moved and established in holding pens on the 174,000-acre Malheur National Wildlife Refuge in southeastern Oregon. Others have been moved to Ruby Lakes National Wildlife Refuge in Nevada, and some to the National Elk Refuge in Jackson Hole, Wyoming. The Red Rock Lakes Refuge has about reached its maximum capacity of these big birds, and for that reason some young are each year transplanted to other suitable areas in the western states so that they may become reestablished over wider ranges. Separation of the flocks is good insurance against disease outbreaks on any one area.

The Ross's goose, a bird that breeds in the Alaskan tundra section and migrates in greatest numbers along the Aleutians into the Asiatic wintering grounds, also moves in rather limited numbers down our own West coast. The Sacramento National Wildlife Refuge in California is the principal known wintering ground in the United States, with numbers varying from 200 to 700 of the birds mingling with the masses of other geese that use this fine area.

The limpkin or "Crying Bird" is another species that is becoming rare and for which this Service is providing much-needed protection on its St. Marks National Wildlife Refuge in Florida. The numbers of this bird in the United States today are unknown, but the species is believed to be threatened because of destruction of its habitat. It is a most interesting bird and its call, once heard, is never forgotten. The Florida crane, another bird which is exceedingly limited in its range, now uses four principal localities, of which three are under Federal protection. One is the famous 330,000-acre Okefenokee National Wildlife Refuge in southern Georgia. The others are the Everglades National Park and the adjoining Everglades National Wildlife Refuge.

Cranes of both the little brown and the large sandhill species use the Federal refuges in their annual migrations from the nesting grounds, some of which are as far north as the Arctic Circle, to the wintering areas in Texas, New Mexico, and other southern states.

Plate 25. "Wood Ducks" given complete protection under the Treaty for many years, is the subject of the 1943 (tenth) duck stamp. Drawn by Walter E. Bohl. Sale 1,169,352.

The sandhills have recently established themselves as breeding birds in a few favorable places in Wisconsin and Michigan. The 91,000-acre Seney National Wildlife Refuge on the Upper Peninsula of Michigan is one of these spots. Several refuges provide suitable habitat for breeding white pelican colonies.

Many conservationists have expressed concern over the status of the long-billed curlew as a breeding bird because of its great size among the shore birds and its apparent need for undisturbed grasslands. Many of the national wildlife refuges provide these ideal conditions. As a consequence, the long-billed curlew has been found on 20 of the refuges, 10 of them in Montana, one in Oregon, one in Idaho, one in Nevada, one in Utah, one in New Mexico, two in Wyoming, two in Nebraska, and one in South Dakota.

Other Refuges

Space does not permit a detailed discussion of all of the refuges in the Federal system. Only a few of them have been selected to serve as illustrations of what can be done to restore waterfowl habitat. Much was made possible by the allocation of emergency funds during depression years, and since that time, duck stamp moneys have provided the finances needed to acquire, improve and maintain these and additional areas. In recent years, some wildlife refuges have been superimposed upon impoundments created by the Tennessee Valley Authority, the Army Engineers, and the Reclamation Service. Through river basin studies now conducted by the Fish and Wildlife Service and the cooperating state fish and game departments on all areas impounded with Federal funds by Federal construction agencies, other refuges will be developed from time to time. Some of these will be managed by the Fish and Wildlife Service, particularly those that fit into the National Refuge Program, while many others will be developed and managed by the state fish and game departments.

Waterfowl refuges represent one of the most important tools of management for this continental resource. Some are large, some are small. Some provide breeding cover and nesting protection. Some furnish sanctuary, food, and rest as the birds move southward in the fall and northward in the spring, or as they are compressed into the southern marshes during the winter. All serve a useful purpose. All are essential. They are all too few to meet the ever-growing needs of the resource.

CHAPTER XII

Refuge Development and Management

*T*HOSE Blue Goose Refuge markers ought to be printed in Chinese, because I'm sure ducks and geese can read English," a hunter once told me, "judging by the way they come out of the north in the fall a mile high until they see those signs. Then they set their brakes and tumble right in. I've never seen anything like it in all my duck hunting experience." I doubt that the lettering on the markers makes any difference to the waterfowl—but what the markers denote certainly does. Behind those large white and blue shields there is protection, safety, food, and shelter for the birds. And that is no accident!!

It takes more than water to attract ducks. Primarily, food and cover determine the character and the quality of the waterfowl concentrations on any given area. For a sanctuary to be fully effective in attracting and holding waterfowl, the environment must continually provide three vital items—food, water, and cover. Of these essentials, food is highly important. Birds must eat in comparative peace and quiet, even as you and I. The refuge cafeterias are so planned and managed. Even trifling variations may spell the difference between a successful area and a poor one.

Great care is taken in the selection of Federal refuges to see that ample and continuing water supplies are available. Records are

Plate 26. The familiar blue goose refuge sign denotes permanent protection for ducks, geese, and other migratory waterfowl.

checked of the drainage basin run-off or of the flow of live streams which supply it: evaporation losses are studied and water table information is secured from the Geological Survey, state water records, and other sources. Physical properties of the waters are determined; soil conditions are analyzed; and, above all, studies are made to determine the limitations on water level management likely to prevail during the growing season.

The real problem in marsh management begins after the dikes and other water structures have been completed and the control gates closed. From that time on, ecological relationships between food plants, water levels, and wildlife must be subjected to continuous observation and study. All animals depend so basically on plants that it is no misstatement to say that waterfowl management, like wildlife management in general, is, in large part, plant management. The abundance or scarcity of ducks and geese in any locality relates very closely to the abundance or scarcity of suitable food and cover plants. Managing the environment, especially the plant resources, is one of the most practical, natural means of assuring waterfowl welfare.

This is not a simple proposition. Maintenance and development of marsh and aquatic flora suitable to waterfowl consists of two correlated phases, one directly productive and the other repressive. To promote the prevalence of desirable plants it is often necessary to do something about well-established but undesirable competitor species. Cattail marshes are fine for muskrats but they often take the place of good duck foods such as smartweeds, wild millet or wild rice. As far as ducks are concerned, cattails are weeds. The same is true of all marsh and aquatic plants that do not clearly "pay their way" in food or cover value. In other words, if these plants are not definite assets they deserve to be classed as liabilities because they occupy space that could be inhabited by more useful vegetation.

Generally, the removal of a particular kind of perennial weed from a marsh flora means the end of its reign in that particular place for several years. Meanwhile, more valuable species of plants have a chance to spread and take over the area. If useful plants do not immediately take possession, artificial introduction of them may be justified. Eventually, of course, the weed species will incline to reestablish itself and dominate again unless control measures are repeated.

Plate 27. Low dams and dikes now store waters on Federal Waterfowl Refuges for the benefit of ducks and geese and fishermen.

On many southern refuges the dense growths of sawgrass and water hyacinth are exceedingly troublesome. Alligator weed is another pest plant, which if permitted to start on a refuge, will completely choke out and eliminate valuable food plants.

Grazing

Grazing is an important management tool in maintaining suitable habitat for many species of wildlife. Natural factors, uncontrolled and unmanaged, may run rampant and choke out useful wildlife food and cover plants. The Valentine and Crescent Lake National Wildlife Refuges in the sandhills of Nebraska furnish good examples of the importance of grazing. At the time when ranching activities were eliminated to remedy the serious overgrazing on portions of refuge lands, some potential nesting cover reverted to a rank growth of weeds and native grasses which the waterfowl shunned. Grazing and hay harvesting were then reinstituted, with specific restrictions on the number of livestock and period of their use. The careful management of these lands has not only helped to maintain a favorable nesting habitat for waterfowl and upland game birds but has contributed very substantially to local economy. Of equal importance is the reduction of a serious fire hazard that would exist in the absence of moderate utilization of the range land and the hay meadows.

On the Sabine National Wildlife Refuge, in the salt marshes of Louisiana, blues, snows, and Canada geese arrive each fall to find an abundance of tender, green vegetative shoots. These are available only because grazing and controlled burning have prevented an impenetrable mass of grass that could not be utilized by any species of wildlife. Long-billed curlews that require short grass meadows for nesting find what they need on the Crescent Lake Refuge in Nebraska where hay is harvested each year after the waterfowl have nested.

While the establishment of refuges may cause some disrupting influences in the local economy, these are compensated to a certain extent by the availability of grazing and haying benefits as well as other economic uses after the refuge requirements are met. For instance, on the 58,000-acre Lower Souris Refuge in North Dakota, with three-fifths of the area in marshes or impoundments, it was possible recently to permit nearby farmers almost 6,000 animal-use months of grazing, while 2,400 tons of hay were harvested. An additional 2,300 acres were farmed under shares.

Controlled Burning

Controlled burning is likewise important in the management of waterfowl habitat. Fire alone is seldom adequate in the elimination of objectionable growths, but when used in combination with water level manipulations, the net results may be very effective in controlling marginal vegetation. It is a common practice on many coastal waterfowl refuges to follow a definite rotation of controlled burning on marsh lands for the purpose of providing browse areas attractive to geese and facilitating the utilization of seeds, tubers, and root stalks by several species of waterfowl. In other sections, controlled burning of uplands is one of the cheapest and most effective means of opening up areas for woodcock and grouse. Controlled burning on many areas serves to maintain desirable stages in natural plant successions and to prevent invasions of brush and other undesirable growths.

Timber

Timber management is also desirable and in some instances a vital part of wildlife management. In the White River bottoms of Arkansas where the 110,000-acre White River National Wildlife Refuge is situated, several million ducks annually frequent the area seeking food and protection. The attraction is the abundance of mast produced by several species of oaks and other nut-producing trees. Commercial cutting operations are regulated so as to encourage these important species. The management of the timber crop is important to wildlife on many of the Federal wildlife refuges.

Farming

Because of the large concentrations of birds which visit the refuges, the natural foods growing in the shallow waters along the margins of lakes and ponds on the refuges are often insufficient to meet their needs. Then corn, barley, soy beans, wheat, oats, and other domestic grains are produced by refuge personnel or by cooperators on a share-cropping basis. Permittees, in lieu of rental, leave the Government's share of grain standing in the fields for the birds or sometimes harvest it and store it for use as supplemental feed in the winter. And do the ducks and geese enjoy it! In 1947, they completely consumed, for example, more than 40,000 bushels on the Tule Lake Refuge and more than 50,000 bushels on the Lower Klamath Refuge, both in northern California.

A farming program, coupled with marsh improvement on the Blackwater Refuge in Eastern Maryland, soon met with an enthusiastic response. In the fall of 1937, 20,000 waterfowl, including 500 Canada geese, used the area. Ten years later, more than 55,000 waterfowl, including 23,000 Canada geese, used the refuge. On one November morning in 1947 the refuge manager found 13,000 Canada geese feeding in one refuge cornfield. And this in spite of the fact that waterfowl along the Atlantic Flyway were then at a very low ebb.

The Wheeler Refuge in northern Alabama on the Wheeler Reservoir, one of the impoundments created by the Tennessee Valley Authority, was set aside largely as an experiment to see what could be done towards improving waterfowl habitat on man-made lakes which fluctuate throughout the season as the needs for power, navigation, flood and mosquito control are met. It has dramatically demonstrated how waterfowl will use an area as soon as it has been improved and developed for them. Formerly, a very small flight of waterfowl used the Tennessee River. Ducks and geese were rare indeed before the impoundments were made, and they still use little of this vast water system with the exception of those areas where special provision has been made to care for them.

More than 100,000 pounds of seeds and root stocks of desirable duck foods have been planted in the bays along the shorelines of the reservoir and annually some 2,000 acres of uplands are seeded to grains and legumes. The migrant and wintering waterfowl now have a supply of food during each season of the year when they are in the vicinity of the refuge, from early fall to late spring. Canada geese showed one of the best responses to management. They increased from 300 in the winter of 1942 to 3,500 in the winter of 1947. Waterfowl in general increased from 8,000 in the fall and winter of 1938 to 78,000 in 1944.

Muskrats

Muskrats are highly important in waterfowl management. These little fellows make their homes on marsh areas across the country, in both fresh and brackish waters. On the waterfowl breeding grounds the muskrat houses are used as nesting sites, particularly by Canada geese. Muskrats may build up large populations within a few seasons and in the absence of regulated trapping, often outstrip the capacity of the marsh to maintain them. The result is depleted habitat having only limited utility to waterfowl. If the area is overtrapped, the marsh growth may develop rapidly to the point where nesting, feeding, and resting by migrant birds is greatly reduced. If muskrat numbers are properly regulated so that marsh growths are kept reasonably open by their feeding activities, ducks and geese will benefit.

On the Malheur Refuge in Oregon muskrat houses now serve the useful purpose of providing nesting sites for Canada geese where, in the early stages of development of the refuge, artificial sites had to be supplied them. Within a few years after the lake was restored old residents in the locality contended that the refuge marshes supported a muskrat population as great as it ever had. From 1941 through 1947, a total of almost 39,000 surplus muskrats had been taken from the refuge, and during that same period there was a tremendous increase in the number of waterfowl using Malheur Refuge. During the fall of 1942 there were about 650,000 ducks and 200,000 geese, or a total of almost 900,000 waterfowl using the refuge, but five years later there were almost $6\frac{1}{2}$ million ducks and 400,000 geese using it. The proper handling of the muskrat population had much to do with the utility of the area for waterfowl. Muskrats taken from this and many other refuges also provide a substantial source of revenue to the Federal government and to the local counties in which the areas are located.

Beavers

Management of beavers is also essential in the waterfowl conservation program. Beaver ponds attract breeding black ducks, teal, mallards, ringnecks, and other waterfowl and provide some of the best nesting habitat to be found in the Lake states and in New England. They also furnish feeding and resting grounds for small groups of waterfowl as they migrate to and from their winter range.

Predators

Predator control must be done on many wildlife refuges in order that the types of wildlife that we are attempting to promote and develop may be given ample protection. Such control consists largely of the elimination of turtles that prey upon young ducks and geese; of bull snakes that rob eggs from the nests of mallards, pintails, shovellers, and other ducks; of skunks that destroy both the nests and the young of waterfowl; of coyotes on some of the western refuges where they often wreak havoc; and of magpies and crows that not only pick open the eggs of nesting waterfowl but also kill the young.

Unusual Products

Wildlife refuges provide returns to the local communities that extend far beyond their primary purposes. Grazing, haying, timbering, woodcutting, and similar products are taken as a matter of course, but there are also others. Stands of bees are permitted on the Camas Refuge in Idaho where they not only provide a source of revenue to the owners but also help to pollinate the alfalfa fields which provide forage for waterfowl and livestock. Blue grass seed is harvested on the Arrowwood Refuge in North Dakota and the Service receives a share of the seed which is then used in wildlife food patch plantings on many other projects. Ice is cut for domestic and commercial use on the Medicine Lake Refuge in Montana during the season when no waterfowl are present on the area. Cranberries are harvested under suitable permits on the Parker River Refuge in Massachusetts. Even cattails produced on the Montezuma Refuge in New York are put to good use as the stems are used by local industries for calking barrels and the caning of chair seats. During the war, the fluff of the cattail heads harvested on the Mud Lake Refuge in Minnesota was utilized in the making of life preservers and buoyant seats for small boats. Blueberries, raspberries, and chokecherries are harvested on many of the Federal areas.

Recreation

Of more importance to the people who live in the surrounding areas are the recreational facilities found on many of the Federal refuges, in places where no similar opportunities are available. In

Plate 28. Brook trout from the waters of a wildlife refuge.

many of the areas of the Dakotas, for example, the only water for fishing, swimming, and picnic uses are located on the Federal refuges. In most instances, tables, fireplaces, and other outdoor facilities are provided. Of great importance to the local folks is the fact that refuge impoundments insure permanent and continuing waters in areas which formerly were often dry during the summer season. Naturally, with so much impounded water on the waterfowl refuges, there is an abundance of good fishing on many Federal refuges which can be made available to the public during the off season when the birds are not concentrated on the refuge. In fact, on some refuges at such places as the White River, Seney, Okefenokee, Lacassine, and Crab Orchard Refuges, fishing is better than average.

Financial Returns to Counties

These development projects are not only a source of recreation but also provide very definite financial benefits to the local communities. Federal law insures that 25 per cent of the proceeds for haying, timber cutting, grazing, and other economic uses is returned in lieu of taxes to the local counties for the benefit of the roads and schools. The total receipts from the Federal refuges amount to around $400,000 per year, of which $100,000 is returned to the local counties.

Upland Game

Federal waterfowl refuges have not only improved duck and goose hunting conditions in nearby marshes but deer, grouse, and other upland forms have also increased. On the White River Refuge in Arkansas, one of the best of the duck wintering grounds, white-tailed deer have increased from some 500 in 1937 to about 6,000 in 1947. Fields have been especially planted for deer, and in February, 1946, 48 white-tails were seen feeding in a 15-acre oat patch. Local citizens credit the refuge with the splendid hunting that they now enjoy in lands bordering the area. A total of 162 deer were taken in the fall of 1947 about the refuge, and according to some of the oldtimers they are more plentiful now than they were 40 years ago. This refuge also supplies a great deal of commercial and sport fishing. Almost a million pounds of fish have been taken in some years

Plate 29. A common scene on National Wildlife Refuges.

by commercial fishermen alone, and several thousand anglers use the refuge each year.

On the Mud Lake Refuge in Minnesota both deer and moose have made spectacular increases. Estimated to number not more than 50 to 70 animals in 1938, deer increased to almost 900 by December of 1947. This latter count was based on animals actually seen during a census by airplane. Moose increased from a single pair known to have been on the area in 1937 to 35, all seen from the air during the aerial survey of December, 1947. The first observation of breeding by this species was made in the spring of 1941 when a cow and twin calves were seen.

Bear and deer are protected on the Seney Refuge in northern Michigan. Both have increased under refuge management. Deer increased from about 1,000 head in 1938 to between 2,000 and 2,500 in 1942. Hunting under state law has been permitted on portions of the refuge and hunters have taken 1,735 deer and 20 bear here since it was opened in the fall of 1938.

Beavers have increased from an estimated 50 pairs in 1936 to well over a thousand in 1943, and the refuge serves as a reservoir

Plate 30a. Deer find the refuges a welcome haven.

Plate 30b. Fine bucks like these spread from the refuge into adjacent areas.

for many streams in its vicinity. Local trappers benefit from the restocking thus affected. Trappers have taken almost 650 beavers from refuge streams since 1939.

Deer hunting, which was considered to be a thing of the past in northern North Dakota prior to the establishment of the Souris Refuges, has again become a favorite sport in that state. From an estimated 100 deer in the wooded strips along the river in 1936, the species increased to approximately 1,350 in 1947. Controlled hunts were held in 1941, 1943, 1945, and 1947, and a total of 2,030 deer were taken.

Trapping and Transplanting

Many state fish and game departments have long used the Federal refuges as a source of stock for trapping and transplanting activities. Texas has taken approximately 6,500 deer, 300 turkeys and 175 raccoons from the Arkansas Refuge. The Wichita Mountains Wildlife Refuge in Oklahoma has supplied deer, raccoons, and turkeys for other coverts. Kentucky Woodlands in western Kentucky

Plate 31. Breeding colonies of Canadas have been established on many Federal refuges.

REFUGE DEVELOPMENT AND MANAGEMENT

has supplied these same three species. On Blackbeard Island in Georgia, deer have been trapped and moved to other portions of the state. Deer and elk have come from the National Bison Range in Montana. Muskrats have been trapped and moved from the Seney Refuge in northern Michigan. In 1941, 8,440 pheasants were live-trapped on the Sand Lake Refuge in South Dakota and stocked on areas as much as 350 miles to the westward. The cost per bird, including transportation, was only 44 cents. In 1947, two bucks and 26 does, trapped on the White River Refuge, were used by the Arkansas Game and Fish Commission to restock depleted areas in other parts of the state.

Establishing New Breeding Colonies

Through the proper manipulation of food and cover and the constant attention afforded by a permanent staff of trained refuge managers on the Federal refuges, considerable progress is being made towards the restoration of breeding colonies of birds that had been eliminated from those areas in the past. After Canada geese have been exterminated from a particular area, it is difficult to restore them, except with the help of man. Geese seem to have a very strong homing instinct both as to wintering grounds and breeding areas, and if the nesting flocks have been killed out of a certain area it is not likely that others will return without some assistance.

That breeding geese can be restored to an area, however, is well illustrated by our experience on the Seney Refuge. A small flock of Canada geese was donated to the Service and placed on the Seney Refuge in 1936. Each year since that time they have nested there and in 1947 a total of 550 young were produced from offsprings of the original birds. These geese migrate southward each fall but return faithfully to Seney and its environs for nesting. In 1942, only 205 migrant Canadas used Seney marshes; by 1943 there were 1,000; in 1945, 3,000.

The Souris Refuge in North Dakota has also been used for the reestablishment of breeding flocks of geese. Pinioned birds were held at the refuge until they nested. The young began returning to the marshes of the Souris for their summer homes and, in the period since 1938, have increased to the point where almost 500 young are produced there annually. The Canada goose now nests on 35 of the national wildlife refuges, many of these flocks having been reestablished through the use of decoy flocks.

The most spectacular build-up of Canada goose concentrations in the history of refuge management has occurred at Crab Orchard Refuge in southern Illinois within the past few years. A large lake was constructed here during the early 30's as a part of the industrial development of the area under the former Resettlement Administration. The Fish and Wildlife Service took over a portion of the project in 1946, and one year later, the entire area was transferred by Federal legislation. A food planting program was initiated in 1946 and a few pinioned geese were placed there to attract the wild migrants to this new addition to the Federal refuge system. That year about 500 honkers dropped in. In 1947, there were 8,000 of these visitors, and in 1948 everyone was astounded to see 30,000 honkers use the fields and waters of Crab Orchard. In addition, 3,000 snows and 5,000 blues made considerable use of the facilities. During spring migration in 1949, some 60,000 geese used the refuge. Folks in this entire neighborhood now thrill to the sight of wheeling flocks of geese and ducks, and hunters have had shooting such as this country never before experienced.

Public Hunting

The great bulk of the Federal wildlife refuges are held as inviolate sanctuaries, but hunting of ducks and geese is permitted on small portions of some of the larger areas, while upland game species such as deer, grouse, and quail are taken on many others. All seasons and limits are prescribed in accordance with applicable state laws. Bow and arrow hunting is becoming increasingly popular in some areas. On the Necedah Refuge in Wisconsin in 1946 archers accounted for 86 deer the first weekend and 129 during the entire season. Similar hunts have been staged on the Arrowwood Refuge in North Dakota and the Blackbeard Island Refuge in Georgia.

Field Trials

Field trials are also being staged on a growing list of Federal areas. Spring field trials are held by the Northwest Oklahoma Sportsmen's Association on the Salt Plains Refuge in Oklahoma. Crab Orchard, in southern Illinois, is now the locale of the National Retriever and Springer Trials, the White River Refuge for the Arkansas

Plate 32. Off to the duck blinds. A goodly acreage on the Bear River Migratory Bird Refuge is open to hunting during the open season on waterfowl.

Field Trial Club, while the Beagle Trials for Alabama are staged on the Wheeler Refuge.

Refuges are Community Assets

National waterfowl refuges are not the dampers on the sport of wildfowling that some folks like to think they are. They serve many purposes, and strange as it may seem, they actually improve the hunting locally while at the same time they protect the birds nationally. When properly managed, refuges furnish many collateral benefits—fishing, picnicking, out-of-door recreation, and enjoyment. In short, they are real community assets.

CHAPTER XIII

Research into Management Problems

*H*UNTING and fishing have developed into big business. Combined, they represent the kingpin of all recreational endeavors of the American public. The demands for more game and more fish are growing constantly and will, undoubtedly, continue to do so as our national population figures rise. We should not confine our thinking alone to the economics of wildlife; certainly the spiritual and aesthetic values it provides oftentimes outweigh the mere sport of hunting and fishing. The out-of-doors and the companionship it affords is reflected in the lives of our people in better citizenship, in greater happiness, in more healthy boys, girls, and older folks, and thus in greater national solidarity and security.

No private industry or business with an investment of even an one-thousandth part of the value of this great natural resource would think of operating with the pittance for fact-finding studies that the wildlife administrators have had. Of all the items that go into the annual budget, research is one of the most important; yet it is the most difficult to justify to the public.

I think a large share of this resistance on the part of budget officers and legislative committees who control the purse strings for the expenditure of public funds can be credited to the research workers themselves. If more of those folks who spend their lives studying and analyzing the intricate problems of Nature would tell the public what they are attempting to accomplish, it would be much better. Their findings are the basis for good administration, yet many of them are unwilling to voice opinions that have not been checked and double-checked to assure the ultimate in scientific

Plate 33. Walter Weber's drawing of "White Fronted Geese" was selected for the duck stamp in 1944. This is a familiar sight to midwest and western hunters. Stamp sale, 1,487,029.

accuracy. If research workers would foresake their microscopes, their autoclaves, and their experiments long enough to give folks some inkling of what they are attempting to accomplish, the whole wildlife management program would gain prestige in the eyes of John Q. Public.

Research that is honestly, objectively, and competently pursued has always paid large dividends. The comforts and blessings of much of our modern life are the direct results of research. Anyone questioning the values of fundamental research might ask himself whether he would be willing to have a major operation without the benefit of an anaesthetic. Our lives are filled with conveniences so commonplace that we have forgotten that research originally brought them to us. The automobile, the railroad, the airplane, the electric light, the radio, the telephone, and the thousand and one conveniences that are ours on every hand all came into being because someone had imagination and persistence. Actually, the industrial research that has been responsible for so much in the way of human progress is less than 50 years old. The first crude chemical laboratory was established only 125 years ago.

We have been slower in developing scientific techniques for the proper utilization and protection of our natural resources than we have in industrial fields. But it is high time that this country considered her renewable resources and discovered more effective means of perpetuating them. This continent, more richly endowed by Providence than any other land on earth, has seen more wasteful use of its natural resources than has ever before occurred in the history of the world for a comparable period of time. Unless our people are willing to follow a wiser course, national disaster awaits us just as certain as the night follows the day. Unforgiving Nature constantly reminds us that as we sow, so shall we reap. Dust storms, eroded and abandoned farm lands, overgrazed and ruined ranges, floods, lowered water tables, drouths, and dried up and polluted streams provide unmistakable warnings. They should convince us that conservation must not be regarded as a sentimental hobby or harmless pastime of impractical nature lovers, daydreamers, duck hunters, fishermen and scientists. Rather, that conservation with necessary research to guide it had better be the urgent business of individuals, society, and of government.

Research in the field of fishery and wildlife management has developed a concept that we cannot successfully separate ducks, deer, grouse, trout, bass, antelope, salmon, and all of the other species of the forests and streams from their environment. Wildlife is a product of the land just as much as grass or trees, and its management is inexorably interwoven with the treatment of soils and waters. Good soil management normally means good wildlife management. Rich, fertile lands produce abundant, vigorous and healthy wild things. Wasted, eroded, gulley-ridden farms produce little more in the way of rabbits, pheasants and quail than they do profitable crops. Dirty, stinking, filth-laden, and erosion-choked streams do not produce sunfish, bass or trout any more than they provide suitable bathing beaches for humans. The starting place in wildlife rehabilitation and management is with the soil and its water supply.

Since the original formation of the Bureau of Fisheries and the Bureau of Biological Survey, the two agencies which now constitute the Fish and Wildlife Service, their major functions have centered on scientific research looking toward the improvement of the species which they were authorized to protect and conserve. Previous chapters have already outlined many of the facts uncovered throughout the years which now serve as a basis for every-day management of waterfowl. The Service promulgates regulations for the hunting of ducks, geese, doves, and other migratory game birds, based upon migration patterns disclosed through years of banding operations. Observations initiated almost seven decades ago have established the nesting grounds and the winter concentration areas for the birds. Research has disclosed a wealth of data on habits and requirements. Yet this recital would be woefully lacking if we did not outline some of the other things that the Service has done to round out the factual basis for managing the nation's supply of waterfowl.

Research can be credited to a considerable degree with continuous sport shooting of waterfowl in this country. Had it not been for the earlier studies on population fluctuations, censuses, migration, distribution, nesting, and all of the other factors that go into the whole question of waterfowl management, there is little doubt that the hunting we know today would have passed out of the picture years ago. Factors unearthed by research workers furnished the basis for regulations which were stringent enough to save breeding stocks in days of dire disaster among the birds. Through the application of research findings to refuge operations, the national wildlife refuges have become effective and productive.

Cooperative Units

One of the important parts of the Federal wildlife research program is the section devoted to the cooperative research and training units financed by the Federal Government, the Wildlife Management Institute, the state fish and game departments, and the Land Grant colleges. One of the crying needs in the early 30's was for technically-trained personnel who could guide the wildlife restoration movement. There were few administrators in those days who had basic training other than in forestry or general zoology. There was no place then where students could get well-rounded

RESEARCH INTO MANAGEMENT PROBLEMS

Plate 34. Horseshoe Lake Canadas winging back north to James Bay nesting grounds.

courses in land, water, and wildlife management. Nine of these research units were originally established, located in Alabama, Iowa, Maine, Connecticut, Ohio, Oregon, Texas, Utah, and Virginia. One of the original nine disbanded in 1938. Their popularity has increased so that since that time others have been added in Oklahoma, Pennsylvania, Missouri, Colorado, Idaho, and Massachusetts.

The units are founded on a most practical basis. Not only do the students secure textbook training but they are also required to conduct research in wildlife management. Studies run all the way from bobwhite quail management to deer and livestock relationships, the habits of mourning doves, and the difficult farmer-sportsmen problems. Many waterfowl projects have been included in the courses of study. Approximately 800 students have received degrees in wildlife management since the establishment of the units and the graduates are now sprinkled liberally throughout the state game and fish departments, the U. S. Fish and Wildlife Service, and other Federal and state conservation units. One reason that the Pittman-Robertson program was able to start off in high gear was due to the fact that there were then available young biologists who had secured excellent basic training at these cooperative teaching and research units.

Lead Poisoning

Some of the earliest studies carried on by the Biological Survey had to do with lead poisoning in waterfowl. Losses from this source have their origin in the large quantities of expended shot that from year to year are scattered in the mud about shooting blinds, on marshes, shallow bays, and lakes. Many birds swallow these shot while searching for food and serious poisoning often results. Dr. Alexander Wetmore, now Secretary of the Smithsonian Institution in Washington, D. C., studied this problem and issued a bulletin in 1919 outlining the extent of the poisoning and describing the symptoms. Sporadic research has continued since that time, and it is hoped that eventually a shot will be produced which will disintegrate in water and thus disappear before birds have a chance to pick it up and swallow it, or if swallowed, will disintegrate so rapidly that it will not be harmful. Considerable progress has been made but it still remains for future workers to develop a harmless pellet.

Botulism

Hundreds of thousands of birds have been fatally stricken with Western duck sickness on many prairie marshes from time immemorial. The malady was originally thought to be caused by alkali poisoning from toxic salts that had leached from the ground and settled in the sump-like basins. E. R. Kalmbach, Biologist, connected with the Biological Survey, and Millard F. Gunderson, of the Department of Agriculture, discovered in 1934 that the trouble was actually due to toxins that resulted from bacteria. The causative organism, *Bacillus botulinus,* is the same as that producing ordinary food poisoning.

Years of study have followed that important discovery and considerable progress has been made. If picked up in time, many of the birds can be saved by placing them in fresh, clean water. Later research developed the fact that ducks not too far gone can be cured by inoculation with antitoxin. Yet, the problem of how to save birds when the number so affected may run into the hundreds of thousands remains another question. Picking up such quantities from

Plate 35. For this pintail, botulism research findings came too late.

widely scattered areas of sticky, shallow mud bottoms is no small job. It requires manpower and expense and, if the outbreak is severe, only a small portion of the sick birds can be saved.

The Service has for years carried on extensive botulism control experiments at the Bear River National Wildlife Refuge located at the edge of the Great Salt Lake in Utah. Water levels in the various pools have been manipulated to determine the effects on the course of the disease and new facts have been brought to light. Since the causative organism which discharges the deadly toxin thrives in the shallow, muddy featheredges, botulism can now be controlled either by drying up or by flooding a given pool. Outbreaks have actually been induced and then stopped on some of the Bear River Refuge pools on an experimental basis. This information has added a great deal to the knowledge of how to handle water properly on the western refuges and has been the basis for changes in the design of control structures.

Plate 36. Light-weight air thrust, flat-bottom boats have been developed to aid in picking up botulism victims.

Predation

Considerable work has been done throughout the years on the destruction of waterfowl and other ground-nesting birds by predators. Bull snakes, skunks, minks, and weasels are destructive in some localities while crows and magpies may take unusual toll in others. As early as 1933, studies in Canada by Service biologists revealed that inordinately dense crow populations were a menace to duck-nesting marshes on some areas near the northern border of agriculture in Saskatchewan and Alberta. Elsewhere, with crows fewer in numbers, the losses were correspondingly less severe. Studies have continued throughout the years to determine how effective the large-scale killing of crows on their roosting grounds in the central prairie states might be in relieving depredations on northern waterfowl breeding grounds.

Research has also developed specific methods for controlling coyotes without being destructive to other forms of life on many of the western prairie refuges. Studies on some of the big-game ranges

Plate 37. Botulism victims being brought to duck hospital for experimental treatment. Bear River Refuge Laboratory, Utah.

of the West have demonstrated quite clearly that coyotes are at times the principal limiting factor in the management of antelope. On those areas the Service has carried on control operations to reduce the predators to the point where more desirable forms of wildlife might survive and increase.

Depredations

Various studies have been carried on throughout the years to develop methods of preventing crop damage by waterfowl and other birds. Duck depredations have been especially severe in some spots in the states of Washington, California, Colorado, Idaho, North Dakota, and to a limited extent, in Nebraska; elsewhere, in limited areas in Arkansas, Louisiana, and Texas, blackbirds and other non-game species have likewise been exceedingly troublesome. Frightening devices such as revolving beams of light, flares, bombs, and harmless explosives have been developed and have aided in the relief of local crop depredation problems. In the worst areas Service biologists have perfected airplane herding techniques which move the birds from private crop lands onto refuge areas where the Federal Government and the states provide food for them.

The local farmers have been educated concerning the habits of the troublesome species. Oftentimes, by making slight changes in agricultural practices such as the earlier harvest of standing grains, bird damage can be completely eliminated without the necessity of attempting to kill the offenders. Lethal methods of control such as the use of poison can be applied in severe cases. The Service, however, always recommends against this type of destruction because its biologists have been able in the vast majority of cases to develop other means which are more effective and less drastic.

Food Habits

Little progress would have been made in the development of the Federal and state waterfowl refuges throughout the country had it not been for the long and painstaking research carried on by trained workers to determine the food habits of the various species of birds. This has been one of the basic, fundamental studies conducted since the inception of the Biological Survey. Hundreds of thousands of individual bird stomachs have been examined. From these examinations wildlife researchers have built up a vast store

RESEARCH INTO MANAGEMENT PROBLEMS

of knowledge about the food requirements of all of the important species of birds in this country. With that knowledge, field biologists were able at the inception of the waterfowl restoration program in 1934 to determine whether suitable food plants were present and whether the soils and water supplies were such that natural foods could be expected to grow on the marshes planned for restoration. Excellent literature has been published on the subject of wildlife food requirements. Outstanding are Government Technical Bulletin 643, "Food Habits of Game Ducks in the United States and Canada," by Dr. A. C. Martin and F. M. Uhler, and Technical Bulletin 634, "Food Habits of North American Diving Ducks," by Dr. Clarence Cottam, both published in 1939. These fine publications are standard reference works for refuge managers in all parts of the country, and are also extremely helpful to the owners of private marshes who wish to improve conditions for ducks and geese.

Plate 38. Sowing rice by airplane on Sacramento Refuge.

Pest Plant Control

Elimination of pest plants on refuges receives much attention by research workers. The methods used must be adapted to the kinds of plants, to local conditions, and to the scale of operations. Special forms of equipment such as under-water mowers and weed cutters have been contrived for the purpose. Elbow grease and a scythe are generally effective for small-scale operations on marsh weeds. For example, success in cattail control varying from about 80 per cent to 100 per cent can be obtained by cutting the plants when the unripened heads are well formed and by a follow-up cutting a month or two later. Simply altering the depth of the prevailing water level sometimes does the trick of drowning out unwanted plants and providing more extensive territory for desirable species. If water levels are amenable to control, this procedure deserves primary consideration.

Another method has come prominently to the front in recent years, partly as a consequence of war-stimulated tests on chemicals. It is the use of chemical herbicides such as 2,4-D. Chemicals as a medium for weed control offer considerable promise but, unfortunately, comparatively little reliable information exists on them as yet. Like the harnessing of atomic energy, their practical value is mainly prospective—a thing of the future. Furthermore, almost nothing is known about the secondary or long-term effects of these agents—particularly how they will affect fish and the small aquatic organisms which serve as food for fish and waterfowl. It is conceivable that chemical sprays intended to control pest plants could accomplish more harm than good.

It is not generally realized that habitat improvement work, including pest plant control, offers greater untapped potentialities for the welfare of waterfowl than any other man-controlled factor. Past invasions by expanding agriculture and industry have greatly reduced the natural habitat formerly used by North American ducks and geese and now our most significant opportunity is to make good use of what remains—improvement of ponds, lakes, marshes, and bays throughout the country so that they may contribute their optimum value to waterfowl. The most serious obstacle to this goal is the lack of reliable "know-how."

Insecticides

Wildlife research must be a continuing function because conditions are ever changing as the pattern of land and water uses are altered. Answers obtained 20 years ago may not suit present-day needs. This is well exemplified by the present necessity of checking the effects of recently discovered insecticides, herbicides, and rodenticides on wildlife. Such products as DDT, 2,4-D, and compound 1080 are exceedingly toxic. Studies on the effects of these substances on fish and wildlife have provided techniques that are designed to avoid excessive damage to desirable species while still remaining effective for the control of those pests for which they are intended. Many of the new insecticides can be harmful in heavy dosages while having little effect in moderate quantities. In the use of poison for controlling rodents, Fish and Wildlife Service biologists have discovered that rodents are apparently color blind while birds are highly sensitive to colors and normally shy away from any foods with unnatural tints. This has led to the use on many of the western ranges of green- and black-dyed poison grains which are entirely effective in killing ground squirrels and prairie dogs but which are completely ignored by seed-eating and insectivorous birds.

Cripple Losses

Biologists are on their way to finding answers to those long-debated questions dealing with after-season cripple losses. Some hunters persist in blazing away at ducks and geese regardless of the range. The result can be one thing only—a lot crippled and few retrieved. Many die, victims of raccoons, skunks, dogs, and other predators, but there is always a certain portion that recover. Checking ducks under fluoroscopes have shown some startling results. At the Spring Lake Refuge, operated by the Illinois Natural History Survey, of 160 ducks examined, 40 per cent had from one to six lead shots embedded in their bodies. Nearly three-fifths of the birds examined were carrying one shot, about one-fifth had two, while two birds had five pellets, and three had six.

These ducks appeared to be quite normal with the exception of a few that were unusually light in weight. The birds were taken from live traps, were able to fly, and apparently were feeding. A

Plate 39. Good retrievers save many a cripple!

few showed healing broken bones. The majority of the pellets were found in the muscular tissues, but many appeared to have penetrated the body cavities.

No one knows the ultimate effect of lead shot in the muscular tissue or the body cavity. Certainly, infection and lessened flight powers may eventually kill the ducks, but it is not at all unlikely that carrying these foreign pellets, particularly of lead, which is known to be toxic, may have definite effects upon the future reproductive capacity of the birds.

Many Questions Still Unanswered

While we can report that definite progress has been made in waterfowl research, there are still many unexplored fields. Here are some examples. We need more data on the Far North breeding ranges, areas that have never been as well covered as they should be. More adequate information is needed on the wintering concentrations south of our own border. We know little of possible winter areas beyond Mexico. Census-taking techniques can always be improved as airplanes, aerial cameras, and other new devices become available. Improved methods of obtaining data on hunter kill and crippling losses must be developed. In fact, we can never sit back and smugly boast that we know all the answers. We will never reach that happy state because new questions arise from year to year as conditions of land and water use alter with human needs.

CHAPTER XIV

State Activities

THE STORY of the early administration of wildlife protective laws by the various state fish and game departments parallels the history of the Federal Government's first efforts to lend a helping hand to the harassed and diminishing waterfowl resource. It is a pretty sorry tale. Regardless of how sincere and zealous the administrator might have been, he was hamstrung and frustrated because of public apathy and lack of sympathy.

The members of the state legislatures were jealous of their authority to make laws governing hunting and fishing. Can't you just hear State Senator Joe Doakes, running for re-election to the State Assembly in 1920 loudly proclaim: "I am proud to tell you sportsmen of this county that I have always fought to see that you shall have your share of the hunting and fishing in this state. I refused to agree to that new-fangled notion that the Fish and Game Department should set the seasons and the bag limits. Why, that could mean rank discrimination against our section of the state! I don't go along with some of these calamity howlers who say our ducks and geese and deer and grouse are over-shot. Game will be with us forever, and so long as I am your Senator, you will get exactly the same treatment that other citizens do." Unfortunately, that kind of thinking still persists in all too many places, and it is not confined alone to members of the state legislatures.

Fish and game departments have always been prize targets for the state political machines, and the greater the wildlife assets, the more tempting has been the exploitation. Generally speaking, these are the only departments that have an independent source of revenue—self-supporting from the sale of licenses and fees. If they

Plate 40. "Shovelers," two males and a female, by Owen J. Gromme, was awarded top place in the drawings submitted for the 1945 stamp. The sale of stamps that year reached 1,725,505.

can be raided they furnish political jobs without the unpopular necessity of levying additional taxes. For years it was taken for granted that with every change of state administration there would be a complete turnover in the Game Department—from cellar to attic. How the change would affect wildlife made no difference.

The butcher or the baker or the barber who had worked the most faithfully for the party or the faction that won the election was appointed local game warden or superintendent of the fish hatchery. He might be a good sportsman and sincerely try to do a job, but if he enforced the laws impartially he was bound to apprehend some powerful member of the political machine—and then he was no longer a warden. If he spent his time arresting only those of the opposite faction, or violators of no influence, or better still, if he made no cases whatsoever—he might retain his badge for as long as four years. In any instance, he knew that he and all of his associates, even down to the office girl and the janitor, would look for other employment as soon as the political complexion changed.

No reflection on butchers or bakers or barbers is intended. They can become excellent game law enforcement agents if they are honest, fearless, able, and willing to undertake the strenuous life of a fish and game protector—provided they are permitted to stay on the job. If they are forced to go back to their earlier trades just about the time they learn what an agent is supposed to do, wildlife is the loser. Butchers, bakers, and barbers make more money at their own trades, anyhow. If you don't believe me, check the records in your own state, and you will probably find that they parallel a recent announcement in the daily papers of one of our large eastern cities. Examinations were being held to select new city employees. A street-sweeper job, requiring an eighth-grade education, paid $4,177 per year; a position as naturalist, to conduct lecture tours for the city park department, required a college degree in biology and paid $2,329.

Happily, the unsavory situation, in so far as permanency of tenure is concerned, is gradually improving with increasing public interest in wildlife conservation. Thinking sportsmen realize the benefits to be derived from retaining in office the efficient public servants who are charged with the responsibility of managing the fish and game resources. Even with the picture brightening somewhat, the mortality rate of top-level state administrators is still one of the most damaging factors in wildlife conservation. All too frequently, the heads of departments must change as the political fortunes vary, or, as sportsmen pressure groups who seek special privileges for their own particular sections of the state demand the scalps of those who resist them.

State Civil Service laws, which provide for the selection of personnel on the basis of qualifications and merit, and the retention of employees as long as they do their work in a satisfactory manner, have done much to stabilize fish and game administration. Many states have commissions made up of men of different political faiths who serve staggered terms of office. This form of organization usually leads to stability of operations. The effectiveness of the department depends to a large extent upon the intelligence and courage of the commission and the degree of authority granted it by the state legislature. Some have almost complete power to manage wildlife, while others are much more limited. The latter are less effective.

To the Federal Aid to Wildlife Restoration Act, better known as the Pittman-Robertson program, goes a great deal of credit for raising the standards of state game administration. Technical personnel can be employed by the states for Federal Aid work only after their qualifications have been approved by the Fish and Wildlife Service. They cannot be removed for reasons ascribed as purely political without the state risking the loss of these Federal funds. The net results have been most salutary. More important, this program has provided the funds for the employment of trained, career workers which were lacking before the receipts from the excise tax on sporting arms and ammunition were set aside for wildlife restoration purposes.

Early State Interest in Waterfowl

Prior to the duck disaster in the early 30's the states took little interest in waterfowl, except to make sure that the annual Federal regulations gave them their share of hunting. That was a perfectly natural attitude, because the Migratory Bird Treaty Act had placed the final authority for waterfowl protection in the Federal Government and the states thereafter concentrated their efforts and funds on the upland species which were their sole responsibilities. License receipts were low, and all too frequently those meager funds were further depleted by legislative diversions to some other activity. Many departments did not have centralized control over the management of the wildlife within the borders of their own states. So, with the Federal Government handling waterfowl regulations, it was not surprising that the states spent their meager resources on deer and quail and pheasants and grouse.

In fact, in those days there was much friction between the state game departments and the Biological Survey, then representing the Federal Government. The public had not yet fully accepted the necessity for the regulation of waterfowl by the Federal Government under the treaty with Great Britain, and this feeling was reflected in the attitude of the state administrators. The present regional associations of state commissioners came into being largely as a result of those early conflicts. The Western Association of Fish and Game Commissioners sprung from dissatisfaction among the states over the handling of waterfowl management, as well as other problems arising from the Federal responsibilities for administration of the vast areas of National forest and public domain in the eleven western states. I first attended one of these meetings in the West in about 1931, and the manner in which the state boys squared away and swung at the Federal bureaucrats made the Joe Louis-Jersey Joe match look like a high school bout. The Midwest Association of Fish, Game and Conservation Commissioners also had its origin in controversy over the waterfowl regulations. The Southeastern Association of Fish and Game Administrators was formed largely to afford a means of securing a better understanding between the states and the Federal Government in the making of mourning dove hunting regulations.

The old-time state administrators who bounced verbal brickbats off the heads of the Federals would not recognize the meetings of those same associations today. There are still heated arguments, to be sure, but the entire tone of the conferences has changed. Now the state representatives review progressive legislation and plan concerted action for its support. Bills sparked by special interest groups are analyzed and plans laid for their defeat. The Western Association and the Midwest Association have each sponsored correlated waterfowl studies by the technical staffs of the respective states to secure data on waterfowl abundance, nesting success, losses, disease, and hunting pressure. The Southeastern Association has been the vehicle for the organization of an extensive study of the mourning dove. All of those studies are correlated by the Fish and Wildlife Service and tied in with similar studies by the biologists of the Federal agency. Information is exchanged freely in a sincere effort to do the best job possible with the funds and manpower available.

The Pittman-Robertson Federal Aid program has contributed substantially to effecting this new Federal-state relationship. It has been the means of developing a growing realization that the job is much greater than can possibly be done with the wholly inadequate total resources available. It is utterly childish for the state and Federal agencies to dissipate their energies and waste their time in jealous bickering. All are working to improve wildlife for the same people—the over-burdened taxpayer.

The Depression Thirties

The stark realization of the critical situation confronting ducks and geese was brought into sharp focus by the drought of the early 30's. Alarmed duck hunters, seeing their sport teetering on the brink of disaster, demanded that ailing waterfowl receive help. Some states pitched in to the limit of their funds and skill, but many still looked to the Federal Government to solve the problem. They lost sight of the fact that duck hunters must purchase state hunting licenses to pursue their sport, and that a duck hunter's dollar in the cash register entitles him to consideration in proportion to total dollars received from all hunters. It was unfortunate that much apathy toward conservation prevailed at a time when leaf-raking and development of memorial parks were common activities financed by numerous Government agencies busily engaged in trying to solve depression unemployment problems.

The finger of criticism need not be pointed at the states alone. The former Biological Survey, prior to "Ding" Darling's regime, was just as bad. It was the Federal agency responsible for the well-being and continued high production of waterfowl. Unfortunately, that agency was neither adequately organized nor financed to cope with a national restoration problem when severe and widespread drought developed and waterfowl populations tobogganed downward. By the time more generous appropriations enabled the Survey to accumulate talent and experience, the lush spending by the depression-cushioning agencies began to come to a close. It was then too late to dramatize the problem and to lend a hand to those states that had funds to finance the sponsor's share of project costs. Nevertheless, the 1934 all-time low in the continental supply of waterfowl registered in the minds of the hunters and administrators, emphasized by tightened regulations with reduced bag limits and severely shortened seasons.

Utah was foremost in taking advantage of Civilian Conservation Corps and Work Progress Administration assistance. With such help the state developed Locomotive Springs, Farmington Bay, Stuart Lake, Clear Lake, and the State Public Shooting Grounds. These high-quality waterfowl units, totaling 30,700 acres, constitute the finest system of public shooting grounds so far developed by any state. All of them produce their quota of young birds each year, thus helping to compensate for the season's harvest. Improved feeding conditions represent a very substantial asset.

The Minnesota Conservation Department acquired and developed its 20,000-acre Thief Lake area in Marshall County. This was an outstanding waterfowl project. One-third of it was made a sanctuary, in accordance with state statutory requirements.

Plate 41. "Redheads" by Robert W. (Bob) Hines, is the theme of the 1946 stamp. Sale 1,725,505.

California acquired and developed the Grey Lodge property in the upper Sacramento River Valley, the Suisun unit on the delta of the Sacramento River at San Francisco Bay, the Los Banos area in the San Joaquin River Valley, and the Imperial Refuge adjacent to the Salton Sea. These four refuges contained 9,500 acres of land and water. Each provided a badly needed sanctuary in a location where protection was lacking and where hunting pressure was severe.

Work such as was done to aid waterfowl in California, Minnesota, and Utah was carried on to a lesser extent in a few other states. The total accomplishment of all states, however, fell far short of the needs of the birds. Nor was the Federal waterfowl restoration program then being carried forward on a large enough scale to take up the slack.

The Federal waterfowl refuges and the few state establishments at that time accomplished a very useful collateral purpose. They vividly demonstrated how the fundamental needs of ducks and geese could be met. Hunters learned that refuge concentrations improved their sport in surrounding fields, ponds, and lakes. Without the attraction of refuges, birds were promptly hazed southward by daily bombardments of gunners. Wildfowlers became more insistent in their demands that an equitable proportion of their hunting license contributions be used to improve and safeguard their favorite sport. Thus we arrived at the tail-end of the 30's, with attention to waterfowl needs that could be classed as too little, but not too late. The stage was set for the performance of the next act in this wildlife restoration show.

CHAPTER XV

The Pittman-Robertson Program

THE Pittman-Robertson Federal Aid in Wildlife Restoration Act of 1937 opened up a new and rich vein of opportunity. As predicted by its sponsors, it turned out to be the most outstanding piece of wildlife legislation enacted since the Migratory Bird Treaty Act of 1918. Through this Act the eleven per cent Federal excise tax on sporting arms and ammunition is deposited in a special fund in the United States Treasury. Annual appropriations from the fund are made by the Congress, which since the fiscal year 1948 has allotted the full amount of the receipts. Appropriated moneys are allotted to the states in the following manner: "One-half in the ratio which the area of each state bears to the total area of all the states and one-half in the ratio which the number of paid hunting license holders in each state . . . bears to the total number of paid hunting license holders of all the states." To benefit the small states, the Act later was amended to provide that no state shall receive more than five per cent nor less than one-half of one per cent of the total allotted to all the states.

The states select suitable wildlife restoration projects and perform the work on them. They are entitled to 75 per cent repayment from the United States for money expended on approved projects. Thus, the states receive three Federal dollars for every state dollar that they invest in these projects.

Actually, the sportsmen themselves shoulder the entire cost of operations: first, through the manufacturer's excise tax which is passed along to purchasers of guns and ammunition; and second, through the purchase of hunting licenses, the receipts from which finance the state contributions of 25 per cent of project costs.

In the first ten-year period from July 1, 1938, to June 30, 1948, collections from the excise tax on sporting arms and ammunition totaled $48,175,429.49. During the same period, the Congress appropriated $34,707,960.88 to defray the costs of Federal participation in this cooperative program. (The $34,707,960.88 appropriated included $11,276,687.37 for the conduct of work during the year beginning July 1, 1948.) The difference between collections and appropriations, or $13,467,468.61, became a backlog of funds to be appropriated later whenever the Congress might decide that the states needed additional wildlife restoration funds.

Plate 42. Dr. Ira N. Gabrielson signs the first Pittman-Robertson Project on July 23, 1938, as Albert M. Day, then Chief of new Federal Aid Division, approves. This project furnished funds to restore waterfowl marshes on Weber River, in Utah.

How did the waterfowl hunters fare in the investment of Pittman-Robertson funds during the first decade of operation? Project records show that 22 per cent of the money went into land acquisition, developmental measures, and surveys and investigations to benefit waterfowl. The rest went into projects designed to aid upland game. Probably you will ask if that was an equitable distribution of the money. Here is the answer. Between July 1, 1938, and June 30, 1947, a nine-year period, the states sold 76,474,438 hunting licenses. Migratory bird hunting stamp sales during that same period totaled 12,558,230. A little figuring shows that almost 17 per cent of the license purchasers were waterfowl hunters. The wildfowlers, therefore, did obtain an equitable share of the Pittman-Robertson dollars.

The Ogden Bay, Utah, project, the first to be submitted, was designed to benefit ducks and geese. The development had a double-barreled objective. Mud flats exposed by the gradual shrinkage of the Great Salt Lake had long been a periodic deathtrap to many thousand of waterfowl because of botulism, locally known as Western duck sickness. Utah's construction work at Ogden Bay differs from conventional design. Ample water to create a duck haven was available from the Weber River which flowed across the gently sloping mud flats enroute to Great Salt Lake, so within the main dike system, with its inlet and outlet spillways, low dikes were erected in the upper parts of the enclosed mud flats. Water turnouts of the multiple type were then installed in these secondary dikes. The result is a combination of irrigation and impoundment controls which increase the production of duck and goose foods on the fertile delta lands, while dikes and channel banks furnish attractive nesting sites. The creation of these secondary water control and diversion units within the main diked enclosures permits the fast manipulation of water—imperative in the checking of botulism. As soon as sick birds appear, the contaminated areas are either drained or water depths are increased. Drainage gets the birds off the land while sluicing in fresh water diffuses the poison to an extent where it is no longer lethal. Outbreaks of botulism among bird concentrations at Ogden Bay have not taken serious tolls since Utah completed its developmental work. The hunter kills, plus birds still periodically lost to the toxin, are far below the losses previously sustained on the undeveloped mud flats.

Plate 43a. Bulrushes were planted along the embankments.

Plate 43b. Alkali flats soon turned into productive breeding marshes.

This 12,000-acre Pittman-Robertson development at Ogden Bay is managed by Utah as a combination waterfowl refuge and public shooting ground. Half of the establishment is open to hunting and half is a waterfowl sanctuary. During four hunting seasons, 18,700 hunters bagged almost 39,000 birds. The Ogden Bay waterfowl layout is a fine duck and goose factory. Nest counts along the banks of a single mile of interior water channel one year disclosed over 200 nests, 17 of which belonged to Canada geese.

Along the Atlantic seaboard, scattered from Maine to Florida, are large expanses of tidal marsh that offer slight attraction for waterfowl. Salt-marsh vegetation produces very little food, and fresh drinking water is lacking. Some of these marshlands offer fine opportunities for creating ponds of fresh water and for producing first-class waterfowl habitat. Transformation of these unattractive waterfowl sites calls for careful studies by biologists and engineers to learn whether the required ingredients are present. If they are, then well-planned and executed construction work will fashion a potent magnet for drawing the air-borne travelers earthward. Use of these duck hotels is not limited to transients. In the northern states breeding birds find attractive nesting sites that have been created for them.

New Jersey's 13,000-acre Tuckahoe Marsh near Atlantic City is an outstanding example of a coastal marsh restoration achievement. In 1941 a Pittman-Robertson project was approved for the development of this marsh. Plans called for trapping fresh water from inflowing creeks along the outer margins of the marsh. Studies disclosed that five fresh-water lakes having a surface area of 1,100 acres could be created, and dikes and spillways were built to impound and stabilize the water.

New Jersey has since expanded the original fresh water impoundments at Tuckahoe. Through a Pittman-Robertson project additional lands bordering the undeveloped marsh have been purchased. Dike and spillway construction will increase the surface area of fresh water to almost 3,000 acres.

Nor has Tuckahoe's middle marsh been neglected. Lack of fresh water prevented the expansion of ponding work into the inner marsh, while at the same time dikes could not be built on the soft mucky soil that underlies those lands. New Jersey did the next best thing.

Plate 44. A nesting Canada Goose resents human interference. Utah Fish and Game Commission Photo.

Using dynamite, a series of quarter-acre ponds was blasted in the interior marsh. These small open-water areas produce some food and also give the birds a place to stop. The Tuckahoe layout is managed by New Jersey as a combination waterfowl refuge and public shooting grounds and hunters now get fine hunting where there was none previously. It rates first class for both purposes.

The example of New Jersey's outstanding Tuckahoe project has stimulated the interest of other Atlantic seaboard states in developing similar coastal marshes. Maryland and Virginia have already selected attractive sites and, through the aid of Pittman-Robertson projects, are going ahead with development work, using the same techniques and procedures developed in New Jersey.

Summer Lake, located in south central Oregon, is another excellent state waterfowl undertaking. Land settlement and the diversion of water flowing to the lake for agricultural uses caused a dwindling of the once broad water surface so that a great expanse of mud flats laid exposed to the north of the remaining open water. Water was available from the Ana River, which heads in a cluster

THE PITTMAN-ROBERTSON PROGRAM 235

of profusely-flowing springs a few miles north of the old lake bed. The 13,000 acres of land needed were owned by the Federal government, the State of Oregon, and private individuals. Required control over the Federally-owned lands was obtained by a public land order issued by the Secretary of the Interior to the Fish and Wildlife Service and then to the state through a cooperative agreement. The state lands were purchased from the State Land Board, and necessary privately-owned lands were bought from their respective owners.

Oregon then proceeded, through a series of Pittman-Robertson projects, to build the required dikes, water control structures, and other features needed to transform worthless stinking alkali flats into a highly productive and attractive waterfowl unit. An average of 6,000 ducks and 1,500 Canada geese are raised here annually. Muskrats abound in the restored marshes, and the annual harvest of these fur-bearers helps to pay for the management of Summer Lake. Half of this establishment is managed as a waterfowl sanctuary while the other half is operated by the state as a public shooting ground. More than 5,000 hunters are now accommodated during each open season,

Plate 45. Baby black ducks on Tuckahoe Refuge, N. J. New Jersey Department of Conservation Photo.

one-fifth of them coming all the way from Portland, 300 miles northwest.

The 30,000-acre Horicon Marsh, in central Wisconsin, is a splendid example of a cooperative restoration project. In 1940 the Wisconsin Conservation Department and the Fish and Wildlife Service entered into an agreement whereby the state would buy and develop the southern part of this dehydrated marsh and the Service would take like action on the northern part. In its original state, Horicon Marsh was a famous waterfowl concentration area and a prolific producer of mink and muskrats. The marsh was drained sometime around 1900 to uncover more agricultural lands, but after being ruined for wildlife it was found that the sour, peaty soils would not grow crops. Another shining example of our lack of water policy!

Wisconsin, through a series of Pittman-Robertson projects, proceeded to buy 10,000 acres of land in the southern part of the marsh, and a subsequent series of developmental projects has since done much to restore it to its pristine attractiveness for wildlife. The state's marsh now yields around 10,000 muskrats per year during the trapping season. Trappers and the state divide the catch equally, with the Conservation Department using its share to help finance the restoration and management of the unit. Half of the state's Horicon

Plate 46. Blasting a one-quarter acre pond in marsh, Tuckahoe, N. J., Pittman-Robertson Refuge. New Jersey Department of Conservation Photo.

Plate 47. Duck hunter's camps at Summer Lake Game Management Area. Oregon State Game Commission Photo.

project is operated as a public shooting ground and half as a refuge. When the Federal portion is completed, with cross dikes controlling the water levels on the upper portion of the marsh, William Aberg and the other Wisconsin stalwarts who crusaded 30 long years for its rebuilding will see the great Horicon bloom again. It is too bad that "Curly" Radke could not have lived to see that day.

Kentucky Reservoir is a 180-mile long link in the chain of manmade lakes created and managed by the Tennessee Valley Authority. To control malaria, a number of shallow water bays along the reservoir margin have been diked off and are pumped out each year. The Tennessee Conservation Department requested the assistance of Fish and Wildlife Service experts in an effort to duplicate here the developments previously made by the Service on the Federal refuges in the Wheeler and Kentucky TVA reservoirs. An agreement was entered into with the Authority to permit the state to farm these dewatered lands, and crops of wheat, oats, corn, peas, and soybeans are now planted annually. In the fall, after the mosquito breeding season has passed, the croplands are again flooded. Thousands of birds now wing their way to these feeding places from the Mississippi River lowlands 75 miles to the westward. And geese are not being neglected as grains and green crops for pasturage are being produced on the reservoir islands. All lands improved by the Tennessee Conservation Department at Kentucky Reservoir are open to hunting, which meets with the enthusiastic approval of the state's duck and goose hunters. The nearby Tennessee National Wildlife Refuge provides needed sanctuary.

238 NORTH AMERICAN WATERFOWL

Plate 48. The Tennessee Conservation Department provides 15 acres of corn and beans and 15 acres of barley for waterfowl on Harmon Creek Island in the Kentucky Reservoir. Photo by Tennessee Conservation Department.

Plate 49. Seven acres of barley and buckwheat on Porter Island area planted by the Tennessee Conservation Department. Photo by Tennessee Conservation Department.

Iowa has stressed acquisition and development of lands for waterfowl in its Pittman-Robertson program. Many relatively small areas have been acquired with a combined total running into several thousands of acres. There are not many opportunities left in a state where 96 per cent of the land is devoted to agriculture, but what the state has acquired and developed to create its present pattern of refuges and public shooting grounds represents high quality. Future acquisitions and developments will be equally good. Iowa's Rice Lake project in Winnebago and Worth Counties is a fine example of a high-dividend-producing restoration investment. The area acquired and developed for this establishment contains 1,900 acres. It is the largest and best known of the Iowa waterfowl units and is managed as a combination waterfowl refuge and public shooting ground.

Missouri purchased 3,050 acres of land along the Grand River in Livingston and Linn Counties for its Fountain Grove waterfowl project. The construction of water-impounding dikes, with included inlet and outlet structures, was completed in 1948. Shallow flooding of the rich bottom lands will produce an abundance of waterfowl foods. This wildfowl establishment is located close to the Swan Lake National Wildlife Refuge which provides needed sanctuary for the birds in that vicinity, so consequently Missouri is operating the entire Fountain Grove project as a public hunting grounds.

Illinois has purchased and developed the 2,400-acre Rice Lake waterfowl management unit in the Illinois River Valley below Peoria. Half of it affords the birds a fine sanctuary, while the balance is a very popular public shooting grounds.

California has bought 8,300 acres of land for its Madeline Plains and Honey Lake waterfowl areas. Grain crops are being produced on part of the lands to provide food for ducks and geese. Half of these lands are open to hunting while on the other half the birds are protected.

New York used part of its Pittman-Robertson income to buy the 2,400-acre Oak Orchard area in Orleans and Genesee Counties. Water-impounding dikes and related control and diversion features have created first-class waterfowl habitat.

Kansas started using its Pittman-Robertson cash in 1941 to buy and develop the 16,800-acre Cheyenne Bottoms in Barton County. Reaching satisfactory price terms with the many-included landowners has been a long-drawn-out process. Cheyenne Bottoms is a fine example of the time-consuming delays that are inherent in purchasing and developing many waterfowl areas. This was the first refuge possibility to be examined by the Fish and Wildlife Service after the passage of the Migratory Bird Conservation Act in 1929. Wide interest influenced the Congress in 1930 to pass a special act to appropriate money to buy the lands, but prolonged and unrewarded oil explorations stiffened asking prices so that lands could not be bought for the amount of money available. Rather than increase the appropriation, the Congress withdrew the money it had made available.

The Fish and Wildlife Service still retained its keen interest in establishing a waterfowl refuge at Cheyenne Bottoms, but because of the price handicap other restoration jobs had to be given priority. Finally, in 1941, with full cooperation of the Fish and Wildlife Service, the Kansas Game Commission decided to buy and develop the Bottoms. With half of the developed marsh and water open to hunting, Kansas sportsmen are now assured of a splendid public hunting area.

Minnesota has acquired 41,000 acres of land for its Roseau waterfowl layout near the Canadian border. When completed, the Roseau project will be managed as a combination waterfowl refuge and public shooting grounds. It holds great promise of becoming an outstanding nesting marsh.

Oregon is busily engaged in buying 10,000 acres of land on Sauvies Island, 20 miles down the Columbia River from Portland. With water levels stabilized and grain crops for ducks and geese produced on the higher lands, this will be an excellent combination refuge and public shooting grounds.

During the first ten years that the Pittman-Robertson program was in operation, 38 states either completed projects to aid ducks and geese or started on them. Never before in game department history has so much attention been given to waterfowl. And, better still, it can be confidently expected that the stress on this line of wildlife restoration will increase rather than diminish. With the present precarious status of the waterfowl populations it is fortunate indeed that the states have chosen to emphasize this vitally important work. Many more refuges to supplement the Federal system are sorely

Plate 50. What's wrong with only 4 ducks a day? Or even a single goose!! Not bad for most folks.

needed, and the production of birds on establishments within the breeding ranges is an added contribution of major importance to the flyways.

Waterfowl Surveys and Investigations

Ten years ago there were few states that employed workers skilled in waterfowl management and marsh ecology. Such technicians were not available, and the states were not then emphasizing the biological aspects of waterfowl work. Since then, course work in the colleges and universities leading to degrees in wildlife management has been providing men for these jobs, and stimulated interest in waterfowl restoration and marsh management has opened up a new field of employment. At present there are more work opportunities for experienced waterfowl and marsh ecologists in the states than there are men to fill them.

In Utah aquatic biologists are studying the effects of various water levels, silting, and salinity on the production of desirable waterfowl food plants. The greatest promise for the control of botulism—a single outbreak of which will kill ducks by the thousands—lies in water-level manipulations. Already changed management practices developed by biologists of the Fish and Wildlife Service and the state have resulted in heart-warming declines in the heavy and tragic losses that occurred in the Great Salt Lake area in the past.

Louisiana, Texas, and other states have learned that there is a close correlation between good muskrat and sound waterfowl management. Too few muskrats permit marshlands to develop into dense vegetative types that are not attractive to waterfowl. Excess numbers result in "eat-outs" in the vegetation, thus creating too many and too large open-water areas. Proper numbers produce ideal dispersal of food plants and open water. These biological facts recently assembled by research workers are being put to practical use. In Louisiana alone, between $100,000 and $150,000 a year is being invested in marsh-improvement work designed to benefit both muskrats and waterfowl. The four million acres of Gulf coast marshes in that state produce approximately six million muskrat pelts per annum. With the sale of those pelts bringing in from $5,000,000 to $8,000,000 to the people of Louisiana each year, it is good management to maintain these marshes in a highly productive state for both muskrats and waterfowl.

Washington, Oregon, California, and British Columbia have banded together to conduct a joint waterfowl study in the Pacific Flyway. Findings from the Pittman-Robertson waterfowl study in Alaska are also included in this important cooperative undertaking. Montana, Idaho, Utah, Colorado, Nevada, Wyoming, and Arizona

Plate 51. The 1947 duck stamp, "Two Snow Geese in Flight," was drawn by Jack Murray. Sale 2,016,819.

have done likewise. These states want to learn more about migration vagaries; large-scale banding of the birds, especially on the breeding grounds, improves that knowledge. These states are also carrying on studies to learn the numbers of ducks and geese that are being produced within their own borders on potholes, farm ponds, stock reservoirs, beaver impoundments, state-managed marshes, rivers, and irrigation canals.

Information assembled on these Pittman-Robertson projects is not frozen within state borders. A single coordinator receives, analyzes, and reproduces all progress reports. These data are then furnished to all game departments participating in the work.

Thirty-eight of the states, plus Alaska and Puerto Rico, have conducted Pittman-Robertson studies on waterfowl. Periodic get-togethers of project personnel to thrash out problems of mutual concern and interest and to explain findings and new techniques useful to other states are an important part of this comprehensive program. Never before have so many states joined together to work on a problem of mutual concern. The results are of immeasurable help to the Fish and Wildlife Service in performing its annual job of recommending appropriate hunting seasons and bag limits. Dissemination of information uncovered by the states also better acquaints waterfowl hunters with population, migration, and production facts. This draws together more closely the administrators of this valuable natural resource and the group that harvests the annual crop.

One of the most encouraging aspects of the entire problem of waterfowl management is this recently demonstrated interest of the state fish and game departments. May it continue and expand to greater accomplishments. With all hands working together in unity toward a common objective, conservation forces may eventually succeed in preserving enough areas to guarantee a permanent, perpetual supply of waterfowl. Let us hope that the hour is not yet too late.

CHAPTER XVI

Waterfowl Conservation in Canada

WHEN Canada came into existence in 1867 by confederation of the present Provinces of New Brunswick, Nova Scotia, Ontario, and Quebec, more than half of the area of the new Dominion consisted of lands entirely under the jurisdiction of the Federal Government. At later dates new provinces were created and additions were made to old provinces out of these lands, but even today a million and a quarter square miles of forest, meadow, and tundra, with all the wildlife therein contained, remain under Dominion control.

References to waterfowl conservation in the early records of the Dominion are scanty. The provinces, even before confederation, had their own regulations regarding wildlife, but there was a tendency to look upon the sparsely-settled Dominion lands as an inexhaustible reservoir of furred and feathered game.

Early Federal Legislation

In 1888 a measure of responsible government was given to the Northwest Territories, which then included four districts corresponding approximately to the present provinces of Alberta and Saskatchewan. The Legislative Assembly of the Northwest Territories made game regulations applying to the districts which now form the southern two-thirds of these provinces, but the regulations could not be considered severely restrictive. In the N.W.T. Game Ordinance of 1893, for instance, no closed season for geese or swans was prescribed; and the closed season for ducks lasted for only about three months—from May 15 to August 23. The use of night-lights and certain other appliances for taking any kind of waterfowl was forbidden; but geese were specifically exempted from provisions which made it illegal to take other waterfowl by the use of narcotics, nets, and traps. Waterfowl eggs might not be taken. Any "traveller, family or other person in a state of actual want" could, however, take birds or eggs to satisfy immediate needs.

The Unorganized Territories Game Preservation Act, a Dominion statute of 1894, applied to those parts of the Northwest Territories not included in the provisional districts of Assiniboia, Alberta, and Saskatchewan, and to the district of Keewatin. This Act established closed seasons for swans, ducks, and geese from January 15 to September 1; forbade the taking of their eggs at all times; and prohibited the use of poison and certain appliances in taking waterfowl. A provision was made that Indians, explorers, surveyors, and travellers in need might take birds and eggs to satisfy their needs.

The above Act was repealed in 1906 and replaced by the Northwest Game Act. The provisions of the Northwest Game Act concerning waterfowl were similar to those in the Unorganized Territories Game Preservation Act, indicating that the measures adopted for waterfowl conservation in 1894 were still considered adequate twelve years later.

The increased tempo of immigration between 1900 and 1914, however, brought about serious changes in conditions in the important waterfowl breeding areas of Western Canada. The combined population of the prairie provinces and the Northwest Territories rose from 440,000 in 1901 to 1,661,000 in 1915; hunting pressure on waterfowl quadrupled, and available breeding areas were seriously reduced by cultivation. It became evident that game legislation applying to Federally-administered territories was gravely inadequate.

In 1917—the year in which the Migratory Birds Convention Act was passed—the Northwest Game Act was amended, and changes of some consequence were made in the waterfowl hunting regulations which had applied to the Northwest Territories for 23 years. The open season for ducks and geese was shortened by a month. Hunting of eider ducks was totally forbidden for five years, and that of swans for ten years. Exemptions were provided for Indians and Eskimos and for explorers and surveyors "to prevent starvation" only.

It is to be noted that since the Migratory Birds Convention Act was passed, regulations for waterfowl protection applying to the whole of Canada have been issued annually under that Act. Such matters as waterfowl open seasons, bag limits, and restrictions on hunting methods in the Northwest Territories are now governed by the Migratory Bird Regulations and not by special legislation for the territories.

The Migratory Birds Convention Act

A political obstacle to the introduction of a coordinated scheme of waterfowl conservation in Canada existed for fifty years after the Dominion came into being. The terms of confederation gave the provinces jurisdiction over wildlife within their borders—an excellent arrangement for mammals, fish, and non-migratory birds. The case of migratory birds, on the other hand, presented difficulties. No province contained within its borders both the breeding grounds and the wintering grounds of a migratory species to an extent sufficient for adequate provincial control of hunting of that species throughout its life cycle. Certain provinces, indeed, were traversed by waterfowl on spring and fall migrations, but saw comparatively few of the birds at other seasons of the year.

Consequently provincial authorities were loath to impose unwelcome hunting restrictions on their citizens, while their neighbors could take an unlimited toll of the same flocks of birds. Although closed seasons were decreed by provincial laws, these may have been determined less by a planned scheme of conservation than by other factors. During the 90's, for example, swans and geese could legally be killed in Ontario (without any bag limits) during eight months of the year. The closed season was from May 1 to September 1, during which period most of the birds were rearing their young in remote and inaccessible districts, and the settlers were fully occupied in field work and had little time for hunting.

That the decrease of game from primitive abundance was causing concern even before the end of the 19th Century is shown by the appointment of a Royal Commission to study game problems in Ontario in 1890. It appears, however, that the problem was still universally looked upon as purely provincial, and there seems to have been no attempt at cooperation between provinces with regard to regulation of migratory bird hunting.

Early in the 20th Century, private individuals began to advocate Federal action regarding migratory bird protection, this movement paralleling similar developments in the United States. Canadian conservationists, prominent among whom was Dr. C. Gordon Hewitt, were alarmed not only by the decrease in game birds but by the threat to agriculture presented by the killing of insectivorous birds, and by the excessive slaughter of sea birds for food, plumage, and sport. The movement was supported by many local fish and game associations, which passed resolutions calling for Federal action.

At first the struggle was difficult. The Dominion Government could not act in this matter without concurrence of all the provinces. Some provinces were strongly opposed to the suggestion of measures which not only would seem to infringe on their rights under the terms of Confederation, but also would be highly unpopular with large bodies of their citizens. After much discussion these provinces were prevailed upon to withdraw their objections, and the way was clear for Dominion action.

The Migratory Birds Treaty between His Majesty and the United States of America, signed in Washington in 1916, was a milestone on the long route of conservation history. In 1917 the Parliament of Canada passed the Migratory Birds Convention Act to implement the Treaty. Thereafter migratory birds held a unique status in Canada in that, while they belonged to the province in which they existed at any moment, legislation governing them was largely a Federal responsibility. This arrangement has worked out happily, the provinces and the Dominion co-operating cordially in regulation of waterfowl hunting and in other conservation measures.

Federal Conservation Machinery

Various agencies of the Dominion Government have in the past carried out duties pertaining to waterfowl conservation. Administration of the Migratory Birds Convention Act, originally a function of the Department of the Interior and later of the National

Parks Bureau of the Department of Mines and Resources, is now carried out by the Dominion Wildlife Service of the latter department, in conjunction with the Royal Canadian Mounted Police and in co-operation with the provincial game authorities.

The Dominion Wildlife Service is charged with co-ordinating Dominion and provincial action in connection wth migratory bird conservation. An essential feature of this co-ordination is the holding of annual conferences, convened by the Minister of Mines and Resources and attended by Wildlife officers of the Dominion and all the provinces. The discussion of problems and the presentation of views and recommendations by experienced administrators and scientists of the Federal and provincial governments ensure that wildlife questions affecting more than one political unit are solved in a manner acceptable to all. The work of organizing these conferences, which now generally take place in Ottawa in the early summer, is handled by the Dominion Wildlife Service.

In addition, the Dominion Wildlife Service is responsible for maintaining that contact with corresponding services in other countries—in particular, the U. S. Fish and Wildlife Service—on which international co-operation in wildlife matters is based.

Other bodies which have been created by the Dominion in the wildlife management field include the Commission of Conservation and the Advisory Board on Wildlife Protection. The former was created in 1909, and was composed of Dominion and provincial representatives and a number of private citizens. Its functions were to study and to recommend concerning conservation and better utilization of Canada's natural resources. It was dissolved in 1922.

The Advisory Board on Wildlife Protection is an inter-departmental committee formed in 1916 for the purpose of co-ordinating action in wildlife matters by various departments of the Dominion Government. It contains scientific specialists and administrative officers drawn from several different departments. By obtaining a symposium of the views and advice of these departmental officials, harmonizing this material, and making it available to departments which initiate measures directly or indirectly affecting wildlife, the Advisory Board on Wildlife Protection performs a valuable service in the cause of conservation.

Until the Dominion Wildlife Service was organized in 1947, the number of salaried wildlife officers employed by the Dominion Government was comparatively small. Apart from the headquarters staff of the Wildlife Division of the National Parks Bureau in Ottawa, there were four field officers, each with a district consisting of one, two or three provinces, and some twenty bird wardens and sanctuary caretakers, several of them employed on a seasonal basis. Under the direction of Hoyes Lloyd, former head of the Wildlife Division, a Dominion-wide organization of voluntary unpaid officers was built up. These Honorary Game Officers, as they are known, are public-spirited citizens who devote time and energy to disseminating information, reporting wildlife observations, and furthering the cause of conservation in many other ways.

There has been continued expansion in the Federal wildlife staff since wildlife management, from being a subsidiary activity of the National Parks Bureau, became in 1947 the basic function of the Dominion Wildlife Service. The number of Dominion Wildlife officers in the field has been increased from four to seven, including one for the new Province of Newfoundland. Other officers, known as Dominion Wildlife Management officers, have been appointed to carry on research in the field, and additional positions as bird warden and sanctuary caretakers have been created in areas where stricter local protection was required. Each summer, a number of seasonal appointments of research assistants to field officers are made; in this way practical training is given to promising advanced students and recent graduates in wildlife management. Corresponding expansion has taken place in the scope and activities of the Ottawa headquarters of the Dominion Wildlife Service and in the number and calibre of headquarters personnel.

Enforcement of the Migratory Birds Convention Act, as well as its administration, was originally a function of the Department of the Interior, but in 1932 responsibility for enforcement was transferred to the Royal Canadian Mounted Police. All members of the R.C.M.P. are game officers *ex officio* under the Act, as are also game and fishery officers of all provinces except (at date of writing, 1949) Nova Scotia, Prince Edward Island, and Newfoundland.

Migratory Bird Sanctuaries

In 1887 the Dominion Government "reserved from sale and

settlement, and set apart as breeding grounds for Wild fowl" some 2,500 acres near Long Lake, Saskatchewan. For more than 30 years this was the only Federal bird sanctuary in Canada. Other areas in Western Canada were "reserved" by the Department of the Interior as a measure of protection for waterfowl breeding grounds, but these areas were never officially recognized as bird sanctuaries.

The Migratory Birds Convention Act, without using the word "sanctuary," provided for special protection for migratory birds in "prescribed areas." The first of such areas set aside under the Act were at Bird Rocks in the Gulf of St. Lawrence, and at Bonaventure Island and Percé Rock, near the Gaspé Peninsula, in 1919. These two sanctuaries are unique in that they were established both under a Federal act and by a statute of the Province of Quebec, symbolizing the common interest of the two governments in the matter.

Since 1919 scarcely a year has passed without one or more migratory bird sanctuaries being created in Canada. As a general rule these sanctuaries are permanent; but a small number of them have been cancelled owing to the fact that they ceased to be of value for migratory bird protection because of changed conditions, such as drying up of marshes.

Control of natural resources in Alberta and Saskatchewan was transferred in 1930 from the Dominion Government to the respective provincial governments. Included in the transferred resources were the majority of migratory bird sanctuaries in those provinces, the provincial governments agreeing to maintain them under the conditions applicable to Federal sanctuaries at the date of transfer. The drought of the 1930's affected the value of several of these sanctuaries, and some of them have, by Dominion-provincial agreement, been cancelled or reduced in area.

On December 31, 1948, there were 74 Federal migratory bird sanctuaries, with a total area of 1,800 square miles. They are located in all provinces (except Manitoba) and in the Northwest Territories, and nearly all the principal types of migratory bird habitat in Canada are represented. Many of these areas are strategically located to protect breeding areas, flyway concentration areas, or wintering grounds. A chain of ten sanctuaries for sea birds (including eider ducks) along the north coast of the Gulf of St. Lawrence provides good examples of the first type. Three large sanctuaries in the James Bay area, where geese congregate on migration, are of the second type; and the third type is well represented by Port Joli Bird Sanctuary in Nova Scotia, where thousands of Canada geese spend the winter.

A number of sanctuaries have salaried caretakers, employed by the Dominion Government where this is justified by the values involved, or employed by conservation-minded individuals or organizations. In addition, the Dominion employs salaried wardens for particular areas not classified as sanctuaries, such as districts where the rare trumpeter swan breeds or spends the winter. All sanctuaries receive general supervision by local detachments of the Royal Canadian Mounted Police.

The National Parks of Canada are primarily recreation areas, but on account of the protection afforded to all wildlife within them they are, in effect, bird sanctuaries. The modest beginning of these parks coincided in time with the creation of the first Federal bird sanctuary in Canada. In 1887 ten square miles were set aside by Act of Parliament as the Rocky Mountains Park, to include the recently-discovered sulphur springs at Banff and the surrounding mountainous area. This park was later extended more than once, and new contiguous parks were created, until today four Rocky Mountains parks—Banff, Jasper, Kootenay, and Yoho—provide a wildlife refuge covering almost 8,000 square miles.

Other national parks in different parts of the Dominion offer a variety of types of habitat, in which almost every one of Canada's surviving native species finds sanctuary in suitable natural conditions. The area of the national parks exceeds 12,000 square miles. Also under Dominion administration is Wood Buffalo Park, a reserved area comprising 17,300 square miles of forests and open plains in northern Alberta and the Northwest Territories, which gives sanctuary to a large herd of buffalo as well as to many other types of wildlife. In addition, each of the provinces has its own parks and areas set aside as bird sanctuaries or game preserves, forming in the aggregate an impressive total.

Provincial Conservation Measures

The history of waterfowl conservation measures by the provinces of Canada is divided into two clearly-defined periods, with the point of separation of these periods marked by the adoption of the Migratory Birds Convention Act in 1917. Since that Act was passed, migratory bird problems have been primarily a responsibility of the Dominion, and waterfowl conservation activities by the provinces have generally been carried out either in co-operation with the Dominion or as subsidiary features of general conservation programs.

Before 1917, all of the provinces had legislation for game protection on their books, but the science of wildlife management was still in an early stage of development and its practice was largely empirical. The comparatively small populations and limited financial resources of the separate provinces, the still-prevalent theory of the inexhaustibility of Canadian wildlife, the incomplete knowledge of migration and other important waterfowl problems, the lack of machinery of co-operation with neighboring provinces and states—all combined to make pre-1917 legislation for waterfowl protection well-intentioned rather than of long-term efficiency. Without these early provincial laws, however, the status of migratory and other game birds might well have reached the danger point before the time was ripe for action on a national and international scale.

Reference has already been made to the game laws passed by the Legislative Assembly of the Northwest Territories before the end of the 19th Century. In 1905 the present provinces of Alberta and Saskatchewan were created out of the prairie region which had previously been known as the Northwest Territories, but it was not until 1930 that their natural resources were transferred from Federal to provincial control. Since that date, however, these provinces have worked vigorously to preserve their standing as one of North America's main waterfowl breeding grounds, both by direct action towards that end and by measures for agricultural rehabilitation, water impoundment, and wildlife conservation in general. As an illustration of what has been accomplished, it may be mentioned that at the beginning of 1949 Saskatchewan administered approximately 100 game preserves and 15 bird sanctuaries, comprising 8,500,000 acres.

British Columbia has a fine record of forward-looking conservation laws and efficient enforcement. In 1913 the Game Protection Act, many of the original provisions of which are still operative, was passed by the British Columbia Legislature. This province took a leading role in prohibiting market hunting and the use of repeating shotguns not permanently altered to reduce the magazine capacity to one cartridge.

Until the close of the 19th Century, the Yukon Territory was subject to the provisions of the Unorganized Territories Game Preservation Act of 1894, a Dominion law. In 1900 the Commissioner of the Yukon Territory was empowered to enact special territorial game regulations. The first Yukon game ordinance was issued in 1901, and provided a closed season for all types of game birds. Later conservation laws have been based on the 1901 ordinance and the Migratory Birds Convention Act, with special provisions arising out of the peculiar climatic and social conditions of this Arctic and sub-Arctic territory.

In 1876 Manitoba instituted a closed season for ducks from May 25 to August 15, and forbade the taking or trafficking in eggs of protected game birds. Spring shooting of ducks was prohibited in 1905. There are no Dominion bird sanctuaries in Manitoba, but in 1949 the province maintained 29 game preserves, covering 7,473 square miles, in 5,049 square miles of which migratory waterfowl received year-round protection. Breeding waterfowl also receive great benefit from the reservation of 3,200 square miles of marsh areas in this province for muskrat propagation.

In the older provinces of Eastern Canada there are records of early legislation for the protection of game birds during the nesting season, with a tendency to cut down spring shooting by making the closed season begin earlier in the year. One of the earliest recorded conservation laws in Canada, passed in Nova Scotia in 1794, protected "blue-winged ducks" (black ducks) from April 1 to August 1. This protection was removed in 1815, but in 1874 Nova Scotia imposed a closed season for all ducks from March 1 to September 1. Ontario (then Upper Canada) instituted in 1860 an April 1—August 1 closed season for waterfowl; in 1887 the closed season for swans and geese was changed to May 1—September 1, and that for ducks and all other waterfowl to January 1—September 1.

In Prince Edward Island, sportsmen's organizations are endeavouring to popularize the hunting of Hungarian partridges as an alternative to waterfowl. The provincial government supported this campaign by setting the opening date for Hungarian partridge in 1948 on the same day the duck season opened, in order to minimize the heavy hunting pressure on ducks at the beginning of the season.

Conservation records in Newfoundland go back to 1845, when an act to protect wildfowl and their eggs during the breeding season (May 10 to September 1) was passed. The dates of the closed season have since then been frequently altered, generally within the aim of increasing protection. In 1948 the open season for migratory waterfowl lasted from October 1 (in some sections, September 16) to October 31. The Migratory Bird Treaty did not apply to Newfoundland before its entry into confederation with Canada in 1949. Newfoundland, however, followed conservation policies similar to those of the Dominion, and at the time of confederation possessed seven wildlife sanctuaries covering 615 square miles.

It is worthy of mention that when the original regulations under the Migratory Birds Convention Act were framed, earlier experience of the provinces with such legislation was of invaluable assistance to the Dominion Government. Annual amendments to these regulations are made only after consultation with the provinces, and frequently important conservation measures undertaken by the Dominion have their origin in suggestions offered by provincial authorities.

Private Conservation Activities

Numerous private individuals and organizations in Canada take part in conservation work either independently or in co-operation with Government agencies.

The best known of these conservation workers was the late Jack Miner, of Kingsville, Ontario. His deep interest in wildlife led him to convert his own property into a sanctuary for migratory waterfowl in 1904, when the idea of wildlife conservation was still comparatively new and unfamiliar to the general public. The waterfowl which learned to frequent his sanctuary became a widely-known tourist attraction. Jack Miner's writings and lectures gained him an international reputation and were of outstanding value in stimulating public interest in conservation. Since his death in 1944 his sanctuary has been maintained by the Jack Miner Migratory Bird Foundation, Inc.

The large and influential organization known as 'Ducks Unlimited, formed in the United States for the express purpose of increasing the supply of game waterfowl, carries on extensive operations for habitat improvement in Canada. These operations are chiefly directed towards the protection of waterfowl breeding grounds in and around the marshes and lakes of the prairie provinces. With the active co-operation of the many Canadian members of the organization and of various Canadian public authorities, Ducks Unlimited has preserved and restored numerous areas of this type which had been threatened by drought and the advance of settlement and cultivation.

A number of other bodies whose interests cover a wide field between aesthetic contemplation of nature and active wildlife conservation exist throughout Canada. Among those most prominent in conservation work may be mentioned the Province of Quebec Society for the Protection of Birds, which sponsored the original suggestions for creation of several now-existing Federal bird sanctuaries. The list of past and present organizations of this nature also includes the Provancher Society of Natural History of Canada, the Canadian Bird Protection Society, the Hamilton Bird Protection Society, and the Federation of Ontario Naturalists.

International Co-operation

The fact that the Canadian-United States border transects the main routes of migration of North American waterfowl makes international co-ordination of conservation efforts a question of primary importance. Much of this co-ordination of effort is based directly on the Migratory Bird Treaty. Hunting regulations, closed seasons, shipment of birds and eggs, establishment of sanctuaries, and permits to take migratory birds for scientific or propagating purposes are matters handled by the Dominion Government within the framework of the Treaty.

There are, however, certain aspects of international collaboration in migratory bird problems which are not specifically mentioned in the Treaty, but which have become of outstanding importance since the Treaty was drafted.

One of these questions is the annual estimation of trends in numbers of North American waterfowl. A generation or two ago this would have been thought impossible, but machinery has now been built up whereby the approximate numbers of each species alive at corresponding dates of successive years may be compared with a fair approach to accuracy. The methods used in Canada to take mid-

WATERFOWL CONSERVATION IN CANADA

winter censuses and to make spring and summer surveys are on the same lines as those used in similar projects in the United States already described. In Canada, the midwinter census has an advantage in that the birds are then relatively few in number and are mostly concentrated in the more accessible southern parts of the country. The census is taken under the direction of officers of the Dominion Wildlife Service, aided by detachments of the Royal Canadian Mounted Police and by private citizens.

Of equal importance are surveys made during the spring and

Plate No. 52. "Buffle Heads Aloft," the migratory bird hunting stamp for 1948, depicts three birds flying into the wind over a rough coastal bay. Drawn by Maynard Reece, the sales amounted to 1,722,677.

summer in the principal waterfowl breeding areas. The systematic taking of such surveys was initiated by the U. S. Fish and Wildlife Service, but in Canada the Dominion Wildlife Service now cooperates to a very large degree in this activity. Some of the provinces and the biological departments of some Canadian universities are also able to provide assistance in these projects. The surveys are commenced early in the season and continued until some time in July. They cover such items as date and volume of spring migration, snow, rain, temperature, volume of spring run-off, possible variations of nesting dates from normal, reproductive success, and predation. This information is required to form complete general and sectional pictures of the status of waterfowl. The surveys are carried as far as possible into the summer, in order that the fall hunting regulations may be drafted on the soundest posible basis.

Bird-banding is another activity which by its nature calls for international co-operation. The Dominion Wildlife Service issues, on behalf of the Minister of Mines and Resources, permits to band birds in Canada and is the Canadian clearing-house for bird-banding information. Its work in this field is closely integrated with that of the U. S. Fish and Wildlife Service, whose official bands are used by all bird-banding permit-holders in Canada except the Jack Miner Migratory Bird Foundation, Inc. The number of Canadians holding bird-banding permits—all of them persons who fulfill strict requirements of integrity and ornithological knowledge—is (in 1949) approximately 200. A number of these are officers of the Dominion Wildlife Service and of scientific institutions, such as universities and museums, and others are private citizens whose assistance in this work is given voluntarily.

The continual and cordial collaboration of the wildlife services of the Canadian and United States Governments which is exemplified in these activities is also expressed in interchange of correspondence and reports, presence of officers of one country as observers or active participants at wildlife conferences convened in the other, discussions of questions of mutual interest, and courtesies and assistance extended to research workers of either country, as occasion may require.

CHAPTER XVII

The Mexican Waterfowl Situation

"Mexico is slaughtering *our* waterfowl." (Italics supplied.)
"Mexican sportsmen are killing off the ducks."
"Revise the Treaty."
"Why should I be restricted to 30 days of duck shooting when the Mexicans have four months?"
"What about that 25-bird bag limit in Mexico while I suffer with four or five?"
"The Mexican slaughter is directly responsible for the short seasons and small bag limits that we in the United States must suffer."
"Ducks are sold on the market in Mexico."
"Armadas that kill thousands of birds at one blast are still in operation."
"It's all the fault of the Fish and Wildlife Service because they permit these outrageous things to continue."
"Let's upset international relations until the Mexicans stop slaughtering *our* waterfowl." (Italics supplied.)

*T*HESE are typical of the remarks that go the rounds of the duck hunting fraternity. They are the kind of statements that come to the desks of the officials of the Fish and Wildlife Service, echoing the sentiments expressed in a few articles which have appeared in national magazines. Resolutions introduced in the 80th session of the Congress, and again in the 81st, called for investigation of the situation in Mexico. Some advocate a revision of the Migratory Bird Treaty which was signed between the Governments of the United States and the United Mexican States in 1937.

These stories have made such an impression upon the waterfowl hunters of the United States that the Mexican duck situation has become a favorite topic of conversation. Although the Fish and Wildlife Service has had trained observers on the Mexican scene for many years—extending back, I might add, more than 15 winters—and even though all of the information obtainable from reliable sources has indicated that some of the news accounts of the duck kill in Mexico were greatly exaggerated, I decided that it was best to investigate the situation at first hand. Accordingly, late in the year of 1948, I spent considerable time in the vicinity of Mexico City, accompanied by Dr. Logan Bennett, Chief of the Branch of Research, and David R. Gascoyne, Assistant Chief of the Branch of Game Management, of the Fish and Wildlife Service.

The killing of waterfowl for sale on the public markets is the root of the evil against which we United States citizens complain. This practice is concentrated around large centers of population, for various obvious reasons. Refrigeration in Mexico, outside of the larger cities, is almost totally absent. Therefore, birds must be killed close enough to market to allow their shipment before they spoil in the normally warm weather which prevails throughout that country. For this reason we confined our activities to the Valley of Mexico surrounding Mexico City. I am convinced, from our discussions with many people who know Mexico thoroughly, that this section is the center of the market hunting evil in that country.

We found "wild duck" advertised freely on the menus of many of the best restaurants in Mexico City. We examined some and found them to be real wild ducks. We witnessed Indian boys standing along the Paseo de la Reforma, the main boulevard of Mexico City, openly selling pintails. We visited the public markets, and although we could not find waterfowl for sale, we did see boxes of mourning doves that had been trapped and were being plucked and offered for sale. The price was seven pesos per dozen. We know that armadas still operate because we heard one fired within a mile or so of where we were duck hunting on Lake Texcoco one morning with a group of very fine Mexican sportsmen.

THE MEXICAN WATERFOWL SITUATION

We made the acquaintance of Mexican officials charged with the responsibility of enforcing the game laws, and they impressed us as being honest, sincere, and conscientious individuals, but woefully lacking in the wherewithal needed to do the kind of enforcement we would expect in the United States. We found that there is little public sentiment to back up their efforts to strictly enforce the waterfowl regulations promulgated under the Treaty. We learned that the salary scales of the public officials of Mexico are exceedingly low for the responsibilities these men carry.

We found hunting pressure that would be absolutely inconceivable to the average American duck hunter—because it is so light. We saw good concentrations of birds consisting in large measure of coots. We were with Mexican officials when they arrested several native Indians and had an opportunity to see the kind of antiquated weapons they used. We learned something of the basic philosophy of the native Indians and Mexicans. To them, waterfowl are considered legitimate items of food wherever and whenever they can be taken, the same as farm boys in many states in this country consider that it is proper to take rabbits for food whenever they are fit to eat.

I am convinced that much of the criticism that we in the United States level at the Mexicans is due to a difference in fundamental philosophy about waterfowl hunting. Here, duck hunting is a cherished sport that is the ultimate in outdoor recreation. Many of our sportsmen spend thousands of dollars each year to provide their own shooting facilities and to protect waterfowl on their private preserves. Every waterfowl hunter in the United States contributes annually toward the management of the resource through the purchase of a Federal duck stamp. Many contribute liberally to restoration projects here and in Canada. We have an inborn love of the sport that has been handed down to each succeeding generation since the United States was first settled by our forefathers.

The duck hunters of this country have long been seasoned to regulation by the Federal Government—to laws which are enforced to the best ability of the Fish and Wildlife Service and cooperating states. U. S. duck hunters have felt restraints and restrictions for more than 30 years, imposed under the 1918 treaty with Great Britain. They expect a job of management; they demand that the resource be preserved so that their youngsters and grandchildren may also enjoy the sport of wildfowling as have the oldsters. Duck hunting is one of our most popular and cherished outdoor sports.

There are few people in this country who cannot afford shotguns and shells with which to participate in hunting. True, they may have difficulty in finding a place where they can build a blind, but at least they are able financially to try their hand whenever the season opens.

In Mexico the situation is entirely different. Game management, as we know it in the United States, is practically absent. Hunting is in the same status there that it was here fifty years ago. The native population, I am convinced, take waterfowl for food pretty much wherever and whenever they can while the birds are south of the border. These folks could never even begin to imagine the quantity and variety of things that are set on the average rural table in the United States. Food, to them, is all-important.

On the other hand, Mexican sportsmen of the type that compare with the ardent duck-hunting enthusiasts in the United States are exceedingly few. We met several very fine sportsmen who, separately, gave us estimates of the total number of real duck hunters around Mexico City. It was surprising. Some said that there might be as many as 100, but others said not over 30 or 40, and it seemed that perhaps 50 would be a good estimate of those who actually participate to any great extent in duck hunting in the vicinity of Mexico City. And these 50, by the way, distribute their hunting to all of the lakes in the entire Valley of Mexico. It is a rich man's sport, and only those with ample income can afford to lease the marshes, employ the guides, and maintain the necessary facilities.

And those sportsmen who can afford marshes for duck hunting have another type of sport that is much more popular. This is the club shooting where live pigeons, live goats, and live turkeys are killed in some of the most difficult shotgun and rifle matches that it has ever been my privilege to witness.

Forestal y de Caza

Perhaps a short review of the organizational set-up in the Mexican government for the protection of game will be of interest. Senor Eulogio de la Garza is the Technical Director of Forestal y de Caza (Department of Forestry and Game), which is a section of the Mexican Department of Agriculture. Senor Luis Macias is the Chief of the Game Department which is a part of Forestry and Game. General Jose Pacheco, an old-line Army officer with 30 years of service in the Mexican Government, is in charge of law enforcement for the entire Forestry and Game Department. This is a wise provision, because the Mexicans have a wholesome respect for an Army uniform, and the enforcement of game laws in Mexico is no child's play, as we learned while we were on this trip.

The Department is not adequately financed, as compared with U. S. standards, selling only about 5,000 licenses per year. License receipts go into the Federal Treasury and the Department is operated from appropriations made by the Mexican Congress. There is no provision for special fees such as the United States Duck Stamp. The entire staff of Forestry and Game officials consists of about 700 people scattered throughout all of Mexico. The number who concentrate upon game activities alone corresponds to the force this country provides to protect our valuable wildlife resources in Alaska. There we have some ten full-time game agents. Mexico has about the same number. In fact, our general game situation in Alaska is in a much more critical condition than is the one in Mexico.

The Valley of Mexico

Senors de la Garza and Macias had arranged a most interesting schedule for us, taking us to the areas of greatest waterfowl concentration and giving us an opportunity to see at first hand the conditions under which they operate. First, we were flown over the entire Valley of Mexico in new Fairchild planes which the Department had recently acquired. These were piloted by capable young officers attached to the Department of Forestry and Game.

We flew northwest from Mexico City across the border of Lake Texcoco and on to Lake Zupango, about an hour's flight distance from Mexico City. As we circled at low level about this lake of some 5,000 or 6,000 acres, I was interested to see the condition of the water. It was dirty and roily and murky as the result of the erosion from the surrounding hills which were denuded of their forests generations ago. Dust from the recent eruption of a nearby volcano had been added to the silt. There was little in the way of waterfowl food, and practically no birds. We could see a few long canoes poled about the marsh by natives wearing large, light-colored straw hats. Made of bleached and whitened wood, those boats stood out in bold relief, even against the murky waters of the lake.

Heading northeastward we flew to Lake Pachuca, marked on the map near the mining town of Pachuca. There was not even water in this lake. It had recently been used as a dumping ground for the tailings from a silver mine. We were told that occasionally in the season of heavy rain there was a little water in the lower sump but at the time we visited it there was neither water nor ducks.

Cutting to the southeastward we flew to Lake Tecocomulco, and here found a very excellent marsh of some 3,000 or 4,000 acres. It was exceedingly small for the vast section of desert country we had been covering, but it looked to be a very welcome place for waterfowl. We saw beneath us great clumps of vegetation, some burning as the smoke spiralled skyward. We found good populations of ducks in rather small flocks well scattered throughout the marsh. We saw cattle grazing among the tall tules and again we saw a few natives conspicuous by their large whitish straw hats as they stood in the long canoes of bleached wood. We saw no hunters' blinds on the entire lake.

As we returned to Mexico City we were impressed with the barrenness of the vast plateau that makes up the Valley of Mexico. The elevation averages about 7,300 feet, and in past centuries the Valley was heavily forested but is now denuded until there is little to see as one flies along except the row upon row of maguay plants. These are similar to the decorative century plants found in parks and gardens in some of our southern states. The juice of the maguay, when fermented, produces pulque, a beer-like drink which is a staple in the diet of the natives of this entire section. It is rich in vitamins and although similar in alcoholic content to our own American beer, it tastes considerably different. And how!! There were a few patches of beans and maize scattered among the rows of maguay, but these were few and far between. We covered some 400 miles before returning to the airport at Mexico City, and had seen most of the principal waterfowl wintering areas in the Valley.

Lake Tultengo

Later, by automobile, we drove to Lake Tultengo, some 100 miles northeast of Mexico City. This was the same lake we had previously circled by air and which my notes listed as Lake Tecocomulco. Our sportsman friend, Raul Girard, who drove us out to the lake, explained that the southern end was Tultengo, because it was once a part of the Hacienda Tultengo, while the northern portion was Tecocomulco, because it was once a part of the Hacienda Tecocomulco. The map maker might use either name. The trip was rough and bumpy, with choking volcanic dust swirling up through the floorboards of the car, as we followed the country road across the rolling maguay-covered hills.

This lake is hunted by sportsmen from Mexico City, with two separate clubs operating, one at either end of the lake. We visited the club which now has its headquarters in the old abandoned Hacienda Tultengo. A portion of the ancient mansion has been converted into a very attractive clubhouse. Here we were outfitted with boats manned by native boys and then poled through the marsh out into the central portion of the lake. We found beautiful stands of tules, some of the largest I have ever seen. The lead boat of our little fleet kicked up a few ducks as we pushed through the waters of the marsh. We found excellent waterfowl foods but saw no hunters as this was not one of the regular hunting days at the club. The members shoot only three or four days each week. We found native boys on scrawny ponies splashing through the shallow marsh looking for the cattle that graze in this area in rather large numbers.

After poling into the middle of the marsh to a stonepile which marked the boundary between Lakes Tultengo and Tecocomulco, we turned about and started back to shore. Suddenly the air was shattered by two loud gun reports immediately ahead. General Pacheco and Senor Macias had their boats pushed through the tules and shortly thereafter they emerged, followed by five native Indians who had been shooting around an area where fire was blazing in the dry tules. They were promptly arrested because they had no licenses and also because they had admitted starting the fire in the marsh so that the birds might be driven out into the open water. These natives had a total of five coots and a collection of the most antiquated weapons I have ever seen. They were ordered to follow us to shore.

As we proceeded another loud gun report was heard at the

Plate No. 53a. Poling through the Lake Tultenga tules.

Plate 53b. Nearby shotgun blasts led General Pechacho to five natives, each with one coot and no license, who were instructed to follow us to shore.

extreme left of the boat in which I was riding with one of the local game enforcement agents. We were immediately pushed back into the tules and came upon a startled Indian boy who had a cinnamon teal and an ancient cannon in his possession. Since he had no license he was relieved of his gun and duck and ordered to follow us to shore. The culprits were told that the fine would be 50 pesos and that they could retrieve their guns by paying their fines on the following Monday at the town of Pachuca, some 50 miles away.

I was very much interested in the weapons and after we reached shore I requested the privilege of firing one. All of the guns, incidentally, had been brought back in the boats still cocked and ready to be discharged. All were crude, muzzle-loading contraptions with hammers some two inches long that pulled back with great difficulty to cocking positions. We were told that they were loaded with black powder and with crude handmade shot, or gravel. Factory-made shotgun shells cost 70 centavos each in Mexico.

I put one of these old blunderbusses to my shoulder and pulled with all of the strength in my trigger finger, but nothing happened. Finally, using my entire hand I was able to discharge the gun and it went off like a young cannon. Black smoke rolled into the air and my shoulder felt as though it had been kicked by a mule. It would have been absolutely impossible for anyone to shoot a bird on the wing with any of the six guns that were confiscated that day. That explained why these hunters had so large a proportion of coots in their possession. The fire in the marsh was set to drive the coots out so the natives, standing in the knee-deep marsh, could shoot them on the water at close range.

As we left Lake Tultengo our road crossed an open drainage ditch recently dug. I inquired of Mr. Girard as to the meaning of this ditch. He shrugged his shoulders and informed me that this is probably the last year there will be any duck hunting on Lake Tultengo because it is being drained to provide more agricultural land. The lake bed is now only about one-third as large as it was a year ago and it is rapidly receding as its life-giving waters disappear down the drainage ditch. Bean and cornfields will soon replace the tules and the ducks. The worries in this part of Mexico, as in so much of our own United States, are not concerned with the slaughter of ducks by hunters but rather with the disappearance of essential habitat.

Lake Texcoco

We had a chance to try our hands at duck hunting in Mexico and found it to be entirely different from anything any of us had experienced in the United States. We had been invited by Senors Julio Estrada and Ramon Llano to join them and learn at first hand the kind of duck hunting they enjoy. We readily accepted but we worried

Plate 54. A blast near my boat led us into the thick tules to a startled Indian boy, who promptly surrendered one cinnamon teal and antiquated muzzle-loading shotgun, and followed us to shore, where he obligingly posed to be photographed.

about outdoor clothing as none of the party had taken rubber boots or heavy clothing to Mexico. We were assured that we would need nothing except a little warm clothing, and that boots were entirely unnecessary. Nevertheless, Milton Lindner, an employee of the Fish and Wildlife Service attached to the U. S. Fisheries Commission in Mexico City, and who served as interpreter as well as "manager"

THE MEXICAN WATERFOWL SITUATION   275

for us, rustled together enough heavy clothing to outfit Dr. Bennett, Roddy Gascoyne, and myself.

Leaving Mexico City at 4:30 in the morning, we drove along the highway skirting the northern edge of Lake Texcoco. Some 30 minutes later, we branched off toward the marsh and stopped at a small shack where we were met in the chilly morning blackness by

Plate 54b. All of the confiscated guns were of the same ancient vintage.

eight or ten Indian boys wrapped in blankets. All wore the large light-colored straw hats such as we had seen from the air and which are typical of native costume in that area. We were provided with shotguns and our hosts urged that we each take 200 shells with us. We told them we had never heard of that kind of shooting, but they insisted that we would need the 200 rounds if we expected to come anywhere near killing the bag limit. Texcoco ducks, we were informed, fly exceedingly high, and we should start shooting at 85 yards, at least.

We settled for 100 shells each and started for the marsh. Following the path which led for a mile or so along the top of a well-constructed dike, we came to boats of the same bleached wood variety that we had previously seen from the air. The party consisted of eight—four Americans and four of the Mexican sportsmen. Each hunter was taken in tow by a native Indian guide and poled out to a blind. These were exceptionally well constructed, nicely painted, with arm rest rails, shell trays, and convenient stools. They were poorly screened, but that seemed to be of minor consequence.

As it began to get light, the birds started moving, and soon the bombardment started. I managed to scratch out a couple of teal, and as each one fell, my native boy, who had been standing in his boat in the open water immediately behind the blind, poled out with great enthusiasm—and racket—to pick up the ducks. He had two dogs that insisted on barking and splashing about. When he returned to the blind he made no effort whatsoever to conceal himself. After having flown over this lake a few days before, looking down on these whitish boats and conspicuous hats, I could well picture how we looked to a duck. Pintails in Mexico act exactly the same as pintails in the United States. The moving birds rose higher and higher and higher; it was not difficult to understand why our Mexican friends had insisted that we take 200 shells each. After a considerable flight of birds, mostly pintails, about all that came by were coots in singles or pairs. We could see great rafts of them sitting out in the water on all sides, but few took wing.

The morning was as bright and clear as any bluebird day I have seen in any marsh anywhere. The sun came up without a cloud in the sky, and by the time it was an hour high I was ready to shed the extra clothing Milt Lindner had found for me. Our guides poled back and forth from one blind to another; the dogs barked, and the few ducks that came over saw to it that they were a long way above us. We used no decoys whatsoever, and the Mexican sportsmen do not indulge in the practice of baiting in front of their blinds.

Sitting idly in the blind on the Texcoco marsh on this bright, shiny morning, the air was suddenly rent by a tremendous B-a-r-o-o-o-m coming from the marsh a mile or so to the eastward. Soon, blue-black powder smoke drifted upward from the tules, and shortly thereafter a large flight of ducks showed against the sunrise sky. We had heard an armada with our own ears, and seen the smoke with our own eyes. Queries of the guides by our sportsmen friends indicated that there might possibly be two armadas on this large lake, and that they were operating spasmodically.

The armada is a combination of guns set to kill large numbers of rafted waterfowl that have been skillfully maneuvered by the natives into the target area which frequently is also baited with grain to keep the birds in front of the set. Since the Treaty, the armada has been outlawed in Mexico, but it is still operated in a few places. It has not been many years since the shooting of armadas and the large single-barreled punt guns of great destructive power were wiped out in the United States. In fact, punt guns still occasionally operate on the eastern shores of Maryland and Virginia. These are all illegal practices but they are exceedingly hard to stamp out. The complete elimination of these murderous contrivances is a most difficult law-enforcement job.

Our party again gathered at the shack about 11 o'clock in the morning, and we totaled 24 birds for the eight men who had been shooting. These consisted mostly of pintails, with one widgeon, one lesser scaup, several green and blue-winged teal, and one green-head

Plate 55. The business end of a Mexican "armada." The trail of black powder crossing the gun barrels will fire them almost simultaneously into massed feeding ducks. Center hammer is fired when rope is pulled by operator at safe distance in rear. Valley of Mexico (1926).

Plate 56. Gathering ducks, after firing "armada." Valley of Mexico, 1926. The Treaty outlawed these destructive devices.

mallard which, we were told, was a rare bird on Lake Texcoco. Our friends told us this was not a very good day but that they seldom managed to get their bag limits of 25 birds; that it took most unusual conditions for them to reach that number even with the customary 200 shells. It was not difficult to understand why duck hunting is tough sport there after watching the manner in which the birds gained altitude that morning in the bright, sunny weather that is usual in the Valley of Mexico.

What impressed us most was the fact that we heard shooting in only two blinds other than the eight occupied by our party. And this on an excellent duck marsh within 30 minutes' distance from Mexico City, where there are two-and-a-half million people!!! I tried to picture a similar marsh near Chicago, or Minneapolis, or Philadelphia, or San Francisco, and my imagination failed to function.

On another day, our party drove westward some 75 miles across beautifully forested mountains to a lake on the headwaters of the Lerma River. Here lies a small lake that is an excellent waterfowl wintering area. After bumping along over rough country roads, we came to the little town of Santiago where the party was held up for what seemed to be an unusually long time. We leisurely visited one of the most ancient churches on the continent, wandered through the market square where people were selling bananas, nuts, oranges, and fresh meats, took a few pictures, and thoroughly enjoyed the scene in this interior village far away from the spots that tourists usually visit. We noted that General Pacheco was not with us.

Eventually, the car in which I was riding took on two extra passengers. Again we bumped along rough country roads down over the foothills toward the lake. Finally we reached an old, abandoned hacienda which was about one-half mile from the edge of the lake. Here we unloaded, and again waited for a considerable length of time. Finally a Mexican soldier walked up to our party and motioned for us to follow. We trailed around the ruins of the hacienda, across a drainage ditch, and down through a cow pasture toward the lake. Soon we came upon a cluster of some fifty people near the edge of the water. In their midst was General Pacheco, flanked by a full squad of soldiers armed to the teeth. Most of the natives looked pretty much on the surly side. We then were told the cause of both the delay in the village and the commotion at the side of the lake. The General, while in Santiago, had received a report that an armada was being set up on the lake. He immediately delayed the party until he could get a squad of soldiers from the local garrison, after which he proceeded to the marsh ahead of us to make sure there would be no shooting. Last year, in attempting to capture an armada and its operators who, incidentally, resented the interference, a gun battle ensued and before the powder smoke cleared away a game warden, a soldier, and six native Indians had lost their lives.

Mexican Waterfowl Kill

The armadas are machines of destruction that have brought on the greatest criticism from American sportsmen. The sooner they can be put out of business the better it will please everyone concerned—Mexican officials, Mexican sportsmen, and our own folks alike. But don't get the idea that the mere passing of a law will do it. If the sale of birds in Mexico City could be eliminated, the armadas would undoubtedly quickly pass out of existence because they are the means by which birds are obtained for the market. I am confident that this will be the principal drive of Mexican officials in the future, aided by the efforts of the sincere and influential sportsmen of Mexico who decry the practice as much as we do. In Mexico, as in the United States, public officials must have the support and backing of public opinion. So far the sale of waterfowl in Mexico City is not looked upon as any great crime. Patience and effort can eventually cure the situation.

The practice of killing with armadas is confined to marshes near centers of large population, such as Mexico City. Warm climate, lack of refrigeration, and lack of transportation make this practice unprofitable elsewhere. Even so, the kill by armadas is relatively small when compared to the total picture. I think the mere fact that birds are killed by these unsportsmenlike methods and are sold in the public markets are what arouses American ire. It is not so much the quantity taken as it is the method.

The situation elsewhere in Mexico is not conducive to a large total take of birds, with the exception of a few spots which are so situated that they can be visited by hunters from the United States. Some of these so-called sportsmen have proven to be slaughterers of the lowest type. They are worse than the armadas.

Dr. George B. Saunders, Service Flyway Biologist, has worked in Mexico continuously during the winter periods for more than ten years. During that time he has traveled by car and by plane, and made numerous contacts with local individuals. He is a thoroughly competent and reliable biologist, and from him we have obtained much information on the situation in Mexico during the past decade. Dr. Saunders estimates that on the Gulf coast in northern Mexico there are not more than 60,000 ducks and possibly 200,000 coots killed annually. The coots often go into commercial use.

On the border region in which are located such cities as Matamoros, Reynosa, Nuevo Laredo, Nogales, Juarez, and others, waterfowl are sold in some restaurants. Here again coots constitute the bulk of the birds offered for sale. Dr. Saunders points out that they would not be sold in the restaurants here any more than in Mexico City were it not for the demand created by American tourists. Also many of the birds that find their way into restaurants are shot by

American hunters who go on duck hunting trips into northern Mexico. Since they cannot legally return across the border with more than the United States limit, the birds are either sold or turned over to the local guides and so find their way into market. He estimates that 125,000 birds are taken annually in the border region.

Along the Pacific coast of Mexico, Dr. Saunders estimates the take to be around 50,000. There are few large towns in this section, so there is very little commercial demand. His estimate of the take of birds in the Valley of Mexico, the area which we visited, does not exceed 175,000. He figures that for the remainder of Mexico, where hunting is extremely light, the total kill will probably not exceed 50,000. The sum total of his estimate for the take of birds in Mexico, therefore, is 460,000.

Other observers, some of whom have been in Mexico for years, give very similar reports. Milton Lindner, who works closely with the Mexican officials, is a little more liberal, setting a figure of some half-million birds as an average take in Mexico throughout a period of years.

In analyzing the best information we have been able to obtain, not only from Fish and Wildlife Service biologists but also from other qualified observers, we must conclude that the take of birds in all of Mexico probably does not exceed five per cent of the total kill on the North American continent each year. Dr. Saunders' comment that more birds are taken in the United States on the opening day of the season than in the entire four months' season in Mexico is difficult to dispute. Even this would be reduced if there were some way to control the murderous hunting indulged in by some American citizens who cross into Mexico where the concentrations of birds is large and where enforcement is practically nil.

Mexican officials and Mexican sportsmen resent articles of the Mexican duck "slaughter" such as occasionally appear in the American press. They point out that Mexico sells 5,000 hunting licenses, compared with 12,000,000 in the United States. We inquired as to the proportion of these 5,000 licensees that might be duck hunters, and were given an estimate of about 20 per cent. No separate waterfowl licenses are sold, but on this basis there would be 1,000 duck hunters in Mexico as compared to our normal number of 1,750,000 to 2,000,000.

Plate 57. Natives on Lake Patzcuaro, Mexico, attempting to spear ruddy duck.

There are undoubtedly more than this in Mexico, considering the native population, but from my own observations and those of others, I do not feel that the native hunters are very damaging to the ducks! As for native hunting, we should go a bit slow in our criticisms. Alaska and Canada both recognize the importance of waterfowl in filling the needs of the native populations for food and clothing. Licenses are not even required where ducks and geese are taken for these purposes. In fact, in Canada, the Dominion Government maintains a number of reservations in the Northwest Territories that are held as exclusive hunting and trapping grounds for native Indians, Eskimos, and half-breeds. No white person is permitted to take wildlife over an area of almost 400,000,000 acres. This is almost double the total area of our National Forests, which comprise the largest public area reserved for any one purpose in the United States. Similar provisions are made in Alaska. The natives of Mexico, in their normal hunting activities, take mostly coots, I am sure. For years, in the United States, we had a daily bag limit of 25 coots, encouraging folks to hunt them and thereby take the pressure off the ducks.

Waterfowl cause depredations in agricultural areas of western Mexico and Mexican officials point out that they have their problems there exactly the same as we have in California, Washington, Oregon, Colorado, North Dakota, and some of our other agricultural areas. When we mention the 4-months' season in Mexico, they logically counter with the fact that we have almost that long in the United States. True, the individual duck hunter in any one state may have only 30 days under the Federal regulations—and that is not much. From a duck's viewpoint, however, the open season in the United States is considerably longer—in 1948, it was 93 days. From the time it crossed the Canadian border on October 8 until it crossed the Rio Grande on January 8, every day of that whole time it was faced with a barrage of guns sticking out of every marsh in the United States except on the refuges. So, it really wasn't a comparison of 30 days of duck hunting in Missouri or Ohio or Maryland against 120 days in Mexico. Rather it was 93 days in the

Plate 58. Tarasco Indian has just thrown double-pointed spear at a coot. Lake Patzcuaro, Mexico.

United States against 120 days in Mexico. Actually, the breadth of these United States is less than the length of Mexico, so the comparison is not nearly as odious as some would have us think. When we total the open seasons in Canada, the United States, and Mexico, we find that waterfowl spend one-half of every year dodging hunters. Is it any wonder that their management is a tough job?

Plate 59. "Goldeneyes"—a male and female winging into a quiet cove to view the courtship antics of two males on the water, was the drawing selected for 1949, the sixteenth stamp of the series. The artist was "Roge" E. Preuss. Sale through March 31, 2,096,252.

We should remember that the hunting of game in Mexico is considered in an entirely different light than it is here. This is probably best exemplified by the fact that we have had a treaty with Mexico for the protection of waterfowl only since 1937. We have had a treaty with Great Britain since 1918. While we have had 31 years of international cooperation with our neighbor to the north, we have had only 12 years with the folks to the south.

Mexican officials are striving to do a good job with wholly inadequate resources and with a public that has an entirely different concept of the place of wildlife in the natural scheme of things than do we. One encouraging result that came out of the recent conference was the development of a plan whereby officials of the Department of Game will visit the United States each year to attend joint meetings of the officials of the Canadian and the United States Governments to discuss waterfowl regulations. We hope to develop a pattern of cooperation that will lead to more uniformity in the making of regulations protecting the waterfowl resource.

I am convinced, however, that the Mexican Government will never follow exactly the same pattern of regulation that we follow here. That is entirely logical when one considers the terrific hunting pressure in the United States as compared with conditions in Mexico. Neither can I conceive that the Canadians will ever follow exactly the same pattern of regulation that we must prescribe here. They, too, have entirely different conditions and situations and feel that they need not follow the same restrictions that the United States hunters must, because Canada produces such a large proportion of the birds that are killed in the United States and has so few hunters in comparison with ours.

Working in close cooperation with officials of both the Canadian and Mexican Governments, I have often noticed an attitude sometimes of tolerant amusement and at other times of impatience bordering upon unbelief at the position of some American sportsmen who seem to feel that every duck and every goose has a U. S. brand plainly marked upon its anatomy. Our neighbors to the north and to the south remind us that Canada and Mexico also have interests in this continental waterfowl population. They remind us rather pointedly of the large proportion of birds that are taken here, as compared with those taken on either side of our international boundaries. They recall that Canada provides production and Mexico furnishes wintering facilities that we do not have. They, somehow, do not subscribe to the claims of some United States duck hunters that "These are *our* ducks." (Italics supplied).

CHAPTER XVIII

Conservation Organizations

IT IS DIFFICULT to decide the order in which we should describe the splendid work of the private conservation organizations. The banding together of sincere and honest conservationists, all seeking to aid in this difficult task of preserving and restoring the nation's fish and wildlife resources, has had an immeasurable effect upon public thinking. These groups have aroused the interest of folks that would otherwise be apathetic toward the plight of the wild things of forest and stream. Each organization has its place in the national conservation picture. All are highly important. In order that we may leave no impression of partiality, it seems best to chronicle the activities of the various organizations in the order of their respective years of service. This we shall attempt to do.

The National Aububon Society

The National Audubon Society heads our list of private organizations because of its age in the conservation movement, dating back to 1886. Many duck hunters have a tendency to criticize the Audubon Society, claiming that the members are complete protectionists and that, if they had their way, the Audubonites would abolish all hunting of waterfowl. That is not the case. The Audubon Society insists only that there be adequate restraints, and it preaches that the waterfowl resource must be given good and sufficient management.

Always on the conservative side, the Audubon Society's philosophy is to think first of the birds and second of the hunters. This philosophy, dictated by long experience, is the only one that can ensure for America the preservation and perpetuation of this great recreational asset. But the National Audubon Society is not solely interested in waterfowl, in fact, not even in birds in general. Its major activities are bound up with the task of conserving all of our natural resources. Its special province is to arouse understanding of the perils of exploitation and of the opportunities which conservation provides. The challenge is not merely to increase or perpetuate the recreational pleasures which our natural resources assure. The real problem is to prevent the human impoverishment which the loss of this heritage would precipitate.

As has heretofore been recorded, the Audubon Society came into being in 1886. Its name was chosen by Dr. George Bird Grinnell who, in the February issue of his magazine, *Forest and Stream,* announced the formation of a society "For the Protection of American birds not used for food." As a boy, Dr. Grinnell had been a student of the widow of John James Audubon, the author of the monumental and authoritative book, *The Birds of America.* The National Association of Audubon Societies, as presently constituted, was incorporated in New York in 1905, with William Dutcher as its first president. Following Mr. Dutcher, the nationally-loved and respected Dr. T. Gilbert Pearson served for many years as its guiding genius.

In the words of John H. Baker, now president of the National Audubon Society:

The people in present-day America, whose interest in conservation hardly dates back longer than a quarter of a century, can scarcely grasp the enormous changes which have taken place since 1886, when the first Audubon Society was formed. In those days, the phrase "conservation of natural resources" had not yet been coined. The principal of protection had not become part of the law of the land. Only a handful of people had the slightest interest in, or knowledge of, the relationship of wise use of natural resources to human welfare.

Sport-shooting, without limit or regulation, and market hunting—a means of livelihood for a considerable number of citizens—had enormously depleted our original huge stock of waterfowl, shore birds and upland game species. Fashion had created a profitable industry in the marketing of wild bird plumage. Egrets, terns, gulls, grebes and other birds had been seriously reduced in numbers or brought to the verge of extinction. The capture and sale of North American songbirds for food and as cage-pets were thriving industries.

If anyone felt that gulls' or terns' eggs were good to eat, he was at liberty to raid a nesting colony and carry off as many pails full of eggs as he chose. It was the fashion for boys to collect bird skins, eggs and nests. They learned to shoot practically in childhood by practicing on robins, doves, sparrows and other birds around their homes.

If an ornithologist of that generation wished to add birds to his collection, he walked to a nearby city park and shot as many warblers in the morning as he could skin in the afternoon. If he wanted warbling vireos, he asked permission of a friend in whose shade trees a pair of warbling vireos were nesting, and shot the birds out of the treetop. Imagine what would happen now to anybody who tried to do that! Public opinion has changed.

* * *

Education is the very heart of our program, the human and social aspect of our work. Since 1911, more than eight million youngsters of school age have enrolled as members of Audubon Junior Clubs, * * * located in every state and in every province of Canada. Each club is organized by a teacher or other adult adviser. Adults have prejudices which must be overcome, but a child only asks to be told. He wants to learn. He craves ideals high enough to give him the thrill of standing on tip-toe to reach them. Therefore, you would expect to find that the Society's greatest educational activity is concerned with America's future.

* * *

We have accepted a self-evident fact, that nature herself is the best teacher. For too many years, it has been common practice to teach biological and zoological subjects on an indoor laboratory basis, with emphasis on the study of skins and skeletons; on anatomy, dissection and taxonomy; on identification and nomenclature. Such instruction is desirable, to be sure, but to see Nature at work in her own laboratory, to observe the way of a robin with its young, to watch a butterfly emerging from its chrysalis, to see a bare rock in the very process of being clothed with soil, or a pond, overgrown with lily pads and pickerelweed, becoming dry land, will thrill and hold ten thousand adherents to every one attracted by the "skin and skeleton" school.

* * *

The Society endeavors to make people aware of the steady stream of life going on around them; of its beauty, richness and complexity and above all, of the fact that they are part of this exciting pageant of events;

that human life is affected in all sorts of ways by every plant that forms a seed, by every woodchuck that makes a burrow, by every dragonfly that emerges from its nymphal case.

But the Audubon Society is not alone confined in its activities to the matter of preaching better conservation and the love of wildlife. It has for years established and maintained wildlife sanctuaries and materially aided in their patrol.

Again quoting Mr. Baker:

Wildlife needs protection over and above the safeguards furnished by law and by majority public opinion. Birds that nest in colonies are susceptible to damaging disturbance, and are especially in need of direct protection. How well protection pays dividends in more birds has been incontrovertibly proved by the abundance of American and snowy egrets throughout the northern states, from Minnesota to Maine. These egrets were on the verge of extinction only 40 years ago.

* * *

Audubon wardens today patrol upwards of two million acres of land and water. Their primary job is to create understanding and good will on the part of people who live in, or visit, the patrolled areas. They are also charged with the direct protection of bird and other wildlife concentrations. They are deputized by Federal and state governments to enforce the wildlife laws and regulations. Patrol work requires men of courage, competence and incorruptibility; men with a genuine interest in protection. These men require boats, cars and other equipment especially fitted for their work and kept in proper operating condition.

In addition to the Rainey Wildlife Refuge in Louisiana, wintering haven for geese and ducks, the Society owns and maintains as sanctuaries the Audubon Nature Center in Connecticut, the Todd Wildlife Sanctuary in Maine, the Roosevelt Memorial Sanctuary in Long Island and, in part, the San Gabriel River Wildlife Sanctuary in southern California. At these places, one may observe the eternal workings of that natural world which is the basis of man's life here on earth; where one may linger while one's spirit is refreshed; where one may come and obtain ideas, take them home and put them to work on one's own land or for one's own community. Here are the familiar birds of town and country, the birds that bring warm cheer and happiness into the everyday lives of men. Such sanctuaries are havens for man and other creatures alike.

The Wildlife Management Institute

The Wildlife Management Institute, organized in its present form in 1946, really had its beginning in 1911, since the Institute has supplanted and supplemented the activities of the pioneer American Game Protective Association which came into being in that year. The parent association was conceived from an early realization that existing stocks of fish and game were not holding up as they should. It was chiefly concerned with the enactment and enforcement of laws designed to relieve the hunting pressures on wildlife. In those days about the only thing that could be conceived as aiding wildlife was to relieve the ever-growing gun pressures.

From this endeavor, the association initiated attempts to stock various game-farm-reared species to augment the natural wild populations. It was one of the forerunners in preaching the gospel of good game management. It accomplished this largely by means of conferences of game breeders and state game officers, calling folks together from all parts of the nation in national conventions. Authorities were invited from the various states, from Canada, and even from Europe to contribute their suggestions and theories in the hope that there might evolve a chance of working out some improvement in the national situation. These early conferences were the forerunners of the present National Wildlife Conference, still sponsored and financed by the Wildlife Management Institute.

Reorganized in 1935 and given the name of the American Wildlife Institute, it expanded and broadened its activities. In 1946 it was again reshuffled and became known as the Wildlife Management Institute. Much credit for the growth and evolution from the American Game Protective Association to the American Wildlife Institute and then to the Wildlife Management Institute goes to the late Honorable Frederick C. Walcott, former United States Senator from the State of Connecticut. Senator Walcott, who was elected in 1929, had long been interested in the activities of the American Game Association and, upon his coming to the Senate, he carried his interest in wildlife conservation into that great body. He was instrumental in having the Senate Special Committee on the Conservation of Wildlife Resources organized and became its first chairman. As such, he aided in the passage of the Duck Stamp Bill. He also pressed for the passage of the first Coordination Act which decreed that the Bureau of Biological Survey and the Bureau of Fisheries should be consulted for advice on how fish and game would be affected by water impoundments which Federal construction agencies proposed to create.

One of the outstanding things accomplished by the Institute is the important part which it has played in the organizing and financing of the Cooperative Wildlife Research Units. These units have been highly successful in furnishing trained men to carry on the wildlife program in the past decade. Although supervised by the Fish and Wildlife Service, the units are financed cooperatively by the Federal Government, the Wildlife Management Institute, the state game and fish departments, and the Land Grant colleges of the participating states. They are truly "cooperative." They have made a splendid contribution to wildlife conservation, and much of the credit properly goes to the Wildlife Management Institute.

The Institute not only cooperates with Federal bureaus, states, and educational institutions, but it also works with individuals who are interested in aiding the wildlife cause. For several years, James F. Bell of Minneapolis has underwritten expenses for intensive waterfowl studies at Delta, Manitoba. In 1939 the Institute came into that picture through the generosity of Mr. Bell, along with the University of Wisconsin and Michigan State College. This important outdoor biological laboratory has obtained valuable information on waterfowl nesting, the identification of species, the sex of ducklings, and the comparative movements of the wild and pen-reared ducks. It has also studied the question of predators and has concentrated on marsh ecology. The work at the Delta station has not only been a splendid contribution on its own but it serves each summer as the headquarters for the staff of Federal and state observers who study waterfowl breeding conditions on the prairie marshes of Canada.

The Institute has also cooperated financially in an important study of the black duck nesting habits and conditions along the Atlantic seaboard and in the maritime provinces of Canada. It has assisted materially in attempts to restore eel-grass in the coastal marshes bordering the New England coast. In this, the Fish and Wildlife Service, Ducks Unlimited, and several of the states have contributed funds, assistance, and cooperation.

But the efforts of the Institute are not confined to improving the situation for waterfowl alone. In Maine, the Kendall Memorial Fellowship was established in 1940, to investigate the possibility of restoring Atlantic salmon in the Dennys River. The Institute has contributed to an Indiana project where the Conservation Department and the University are working on techniques to improve fishing waters for bass. Also, the Institute underwrote the preparation and publication of a life history of the striped bass which has been worrying sportsmen and conservationists alike along the entire Atlantic Coast. Following this study, a campaign was launched—and quite successfully—for the passage of more uniform state laws in the states extending from Maine to North Carolina.

The Wildlife Management Institute has published a series of excellent books on various wildlife subjects. One of the outstanding books is *The Ducks, Geese and Swans of North America,* by F. H. Kortright, of Toronto, Canada, illustrated by color plates which do the difficult job of reproducing eclipse and fledgling plumage as well as that of the mature birds. Other works include a series on predators by Stanley P. Young, of the U. S. Fish and Wildlife Service. This organization is gradually accumulating an excellent bookshelf of authentic works on the great out-of-doors.

The American Wildlife Institute, in the first year of its organization, which coincided with the first year of "Ding" Darling's regime as Chief of the Bureau of Biological Survey (he had much to do with its organization), sponsored the first North American Wildlife Conference. It was assembled at the invitation of the late President Franklin D. Roosevelt and was a huge success, as have been all of its successors. Held annually, these affairs have developed into a monumental institution for folks interested in wildlife management. Discussion panels are held where technicians, administrators, sportsmen, and everyone interested may review the problems of the day and report progress both in research and in management. Controversies and differences of opinions—and there have been some good ones—are aired and dissected at these public forums. The proceedings are published each year and represent the best reference work obtainable anywhere on the progress of conservation efforts in the United States, Canada, and Mexico.

Since the organization of the Wildlife Management Institute in 1946, with Dr. Ira N. Gabrielson, former Director of the U. S. Fish and Wildlife Service, as its president, the organization has expanded its activities into still broader fields. Through the addition of a capable field staff, the Institute's efforts and influence are being spread more generally throughout the United States. The field representatives assist the state fish and game departments, the U. S. Fish and Wildlife Service, the Forest Service, the Soil Conservation Service, and numerous private organizations in the moulding of public opinion to do a better job of fish and wildlife management. Dr. Gabrielson has personally studied the operation and organizational structures of several state fish and game departments and has rendered his unbiased report to the state commissions. This service has been exceedingly popular to sportsmen and administrators alike.

The Izaak Walton League of America

This organization was born in large part as the aftermath of a thought-provoking article published in 1921 by Emerson Hough. He was active during that period when thinking men and women were first awakening to the sorry plight of one of America's greatest natural resources—the out-of-doors. Let us review some of the highlights of the clarion call which he titled, "It is Time to Call a Halt."

Scrambling for the last remnants of our great outdoor heritage, we have been so busy as to be blind. Now the truth comes home. Now for the first time a sudden consternation comes to the soul of every thinking man who ever has loved this America of ours.

It is time to call a halt * * *.

Of the alleged true sportsmen of the country, those who use rod and gun, not 10 per cent have practiced the creed which hypocritically they profess. Claiming self-denial, we practice self-indulgence. Which shall first cast a stone? And yet, my brothers, it is time to call a halt.

Never has transportation been so cheap, so rapid. There is no longer any wilderness. Betrayed by its guardians, forgotten by its friends, the old America is gone, and gone forever. Never again shall we have more than fragments. *If even these be dear, then surely it is time to call a halt!*

Those courageous words, written in 1921, soon burned themselves into the consciousness of thoughtful men and women who were beginning to realize that things were not going too well with the country's woods and waters and the wildlife that depended upon them. That article succeeded in crystalizing fears which had been growing daily in the minds of members of a steadily-increasing army of conservationists. It was a cry which rallied many folks to the battle to halt the decline in America's outdoor heritage. It was, in fact, the birth announcement of the Izaak Walton League of America, which came into being on January 14, 1922.

In order that the record may be as unbiased as possible, since I and many of my associates have been enthusiastic members of the Izaak Walton League almost since its inception, it seems best to call upon Harold Titus, the "Old Warden" of *Field and Stream* and long-time member of the Michigan Conservation Commission, to give his impressions after having made a factual study of the principals and the purposes of the League.

He tells us:

There's never been any thing like it. Never before has a group of outdoor men and women, organized on a national basis, fought so consistently and so valiantly for ends which were wholly unselfish. And possibly the most remarkable factor in the League's entire history is that today, after this long service, it is neither weak nor weary; it is stronger, healthier, more vigorous and potent than it ever was before, with no debts due, no apologies due, no embarrassing obligations assumed, and with its code and objectives as bright and shining as they were the day they were drawn up.

For thousands who read these pages the story of the Waltonians is familiar; but for thousands of others the League is one of those taken-for-granted influences in national life, the detail of which is foggy. A number of states have never organized a single chapter; countless hard-working conservationists, because of their location, have failed to come even in remote contact with the inner workings of this remarkable band.

To get the feel of the times which gave background to Hough's great rallying cry, it is necessary to look back briefly to the early twenties, when the country had just reached the limit of its geographical expansion, when physical pioneering was about done, when the awareness that even the most abundant natural resources can be exhausted had only commenced to penetrate the understanding of the most alert. The United States had helped win a great war, had shaken off a depression that for months promised to be disastrous, and was embarked on a boom to set a record for all booms. That anything could be wrong anywhere was almost a treasonable thought in many quarters.

Things were wrong, however. Many things. In many localities. All of them, so far as this recital is concerned, having to do with the outdoors. The passenger pigeon had been gone for a quarter century. The buffalo persisted in pathetic, well-nursed remnants. No one gave those species much thought, though, because they and agricultural men could not have survived together. But the wild-turkey range was shrinking. Quail were disappearing as clean farming practice spread. Antelope were a rarity on their old ranges. Many lakes had ceased to yield fish to angling effort.

Thousands of miles of once-clear rivers had become industrial sewers. Timber—in virgin stands—was a memory in New England and the Lakes states, and erosion was getting in its first noticeable licks.

But we still had forests; remnants of them in older timber-producing areas, vast untouched stands in the South and Far West. We still had fishing. If some of our native upland birds were done for in favorite coverts, we had the tough and challenging ringneck coming along to replace them. Big game still hung on in a good many places. Drainage had ruined some duck marshes, but there were plenty of good ones still left.

And we had the automobile, didn't we? If fish and game were scarce in the home county, good roads made it easy to get into the next and the next. As a matter of fact, for a lot of us fishing and hunting were better than they had been in years because of those motor cars and highways. We could get to where wire fences hadn't discouraged quail, or into new grouse territory, or to where logging and fires had not yet blanked out deer herds, to where trout and bass thrived under almost primitive conditions, and we could get there as easy as falling off a log. What, then, was there to worry about?

Well, there was plenty. Oh, brother, there was everything to worry about! So many of us never considered that the very ease of access which blinded us to what was in the cards, the very triumph of man over time and distance, marked the end of the outdoors that we had always had for the taking. That was the situation which led to the formation of the Izaak Walton League of America, back in January, 1922, when fifty-four far-sighted and apprehensive anglers met in Chicago to talk over the alarming state of fishing in familiar waters and to plan some campaign to halt that blighting trend. * * *

Before those fifty-four minute-men of conservation adjourned, the Izaak Walton League of America was formed to defend the nation's woods, waters and wildlife.

One of the first major achievements than can be credited to the Izaak Walton League was its influence in the establishment of the 300,000-acre Upper Mississippi Refuge for fish and wildlife. Through its efforts the first appropriation by the Congress for the establishment and management of that national wildlife refuge was secured. Soon the League championed the cause of starving elk in the Jackson Hole country of Wyoming, secured funds, and made the initial purchase of lands which were turned over to the Federal Government to become the nucleus for the present highly important National Elk Refuge, now taking care of the wintering needs of the largest single herd of elk in the United States.

In 1928 the League lent its efforts towards obtaining funds for the establishment of the Bear River National Wildlife Refuge in Utah. This had long been a section were botulism was rampant and where many birds each year died as a result. The League aided materially in securing appropriations for a ten-year program to establish the nation-wide system of migratory waterfowl refuges. The League participated in a successful campaign to restore the once-famous Horicon Marsh in Wisconsin as a waterfowl breeding and resting area. This refuge has now become a reality, financed in part by Pittman-Robertson moneys and in part by appropriations from Federal duck stamp funds. The League aided in enacting legislation which authorized appropriations for the establishment of the Federal Wildlife Refuge at Cheyenne Bottoms in Kansas. The Federal refuge never became a reality because of lack of funds, but it is now being developed as a Pittman-Robertson project by the State of Kansas.

One of the outstanding efforts of the League has been its fight against stream pollution. It has always been in the forefront in attempting to clean up America's waterways and streams and to prevent future pollution from industrial and municipal wastes. When it comes to the question of fighting pollution, the Izaak Walton League has become synonomous with clean and pure streams.

The League has aided the U. S. Forest Service in many worthwhile projects. It has fought the construction of high dams on many of the finest fishing streams in America, loudly proclaiming that some areas should be left in the status that Nature intended. The "Ikes" have contributed powerful assistance in the following projects: saving Cumberland Falls in Kentucky from commercial exploitation and the Black Water Falls in Virginia from destruction by power dams and diversions; pushing for the passage of legislation authorizing the creation of the National Park on Isle Royale in Lake Superior; fighting for the sealing of abandoned coal mines in Pennsylvania, Virginia, and other eastern states; establishing adequate fishways on Bonneville Dam; and the revising of grazing leases on the Public Domain to insure that these lands and waters could not be retained by the lessees as private hunting and fishing sanctuaries.

In short, the Izaak Walton League of America, "Defender of Woods, Waters and Wildlife," has well lived up to this slogan. It has a following of thousands of sincere and conscientious conservationists throughout the country who are willing to fight to the last ditch to preserve some of the natural heritage of America. By its actions it has effectively accomplished many of the principles proclaimed in Emerson Hough's battle cry, "It is Time to Call a Halt."

Ducks Unlimited

Ducks Unlimited was conceived and organized in a manner that is typical of American progress. The history of the conservation movement demonstrates that few of us folks get excited about a species that is passing into oblivion until it is almost gone. In some cases we wait until the horse has been stolen before the barn door is closed, but in others we wait until he is about half-way out of the corral before we challenge the horse thieves. The great drought of the prairie states in the early 30's, when many ardent duck hunters realized that the curtain was just about to drop on the sport of wildfowling, was really the beginning of the Ducks Unlimited movement. The horse, then, was at least half stolen.

The present organization was fathered by an earlier group known as "More Game Birds in America." This parent organization, consisting of ardent duck hunters with considerable means, met as a committee in Detroit, Michigan, in October, 1931, to discuss some practical plan for the future of waterfowl management in Canada and the United States. After consultation with officials of the Canadian government and the U. S. Biological Survey, they issued a small booklet entitled, "More Waterfowl by Assisting Nature."

In this booklet they recommended an International Migratory Waterfowl Agency which would be financed by funds to be raised from a "cent-a-shell" tax, supplemented by such governmental appropriations as might be obtained, to promote production with the highest efficiency by acquiring breeding grounds, either in the United States or Canada. They planned that there should be personnel which they would have called "Game Bird Management Forces" to control water levels and provide ample supplies of food and cover, to control natural enemies where necessary, to prevent fires and stop unauthorized grazing, and to eliminate shooting on the breeding grounds. Refuges were to be established by the agency for the use of wildfowl on northern and southern flights, these to be coordinated with a system of concentration areas for winter use. They recommended that the international agency should be organized so that it would function respectively under the supervision and control of the existing departments of the Canadian and United States Governments which were concerned with the administration of laws relating to migratory waterfowl.

The original and announced purpose and intention of the "More Game Birds in America" foundation had, at its inception, the enthusiastic support of both the Biological Survey and the Canadian Government, but soon contention began to dog the path of the organization. Suspicion arose among the Federal and state administrators who were charged with responsibility for making waterfowl regulations and enforcing them that the sponsors of "More Game Birds in America" were more interested in securing hunting regulations to their own liking than they were in following out the high-principled creed that they had announced. Although "More Game Birds in America" as an organization made many friends, at the same time, it accumulated a formidable array of sceptics and opponents.

In 1937, after breeding ground restoration had gotten off to a good start in the United States as a result of the emergency funds made available by the Congress and the duck stamp revenues, some officers of "More Game Birds in America" saw the need for similar activity north of the border. As set up, the foundation could not function there, so "More Game Birds in America" passed out of the picture and Ducks Unlimited was incorporated under the wing of the earlier organization.

Through many vicissitudes, some of which were due to earlier difficulties, and with some later controversy with Federal officials principally over what those individuals considered to be unnecessarily optimistic propaganda of waterfowl supply during years when drouth conditions were again striking at the heart of the breeding areas, Ducks Unlimited has taken a firm place in the hearts of duck hunters scattered from the Atlantic to the Pacific coasts. Its members, many of whom are wealthy sportsmen, contribute substantial funds to do the job of restoring and improving breeding grounds in Canada, particularly in the prairie provinces which are so important in the waterfowl management program for the entire continent.

The philosophy of Ducks Unlimited is sound. Canada, with the principal breeding ground for North America's ducks, has never had the men nor the money to tackle the job of nesting ground restoration. Here in the United States, where the great bulk of the Canadian duck crop is harvested, are the hunters who are able to finance efforts to rectify in some measure what agricultural expansion has done to those northern prairie sloughs and lakes. Ducks Unlimited has attempted to use the private funds of the United States duck hunters to do a job similar to that done in the United States with Federal duck stamp dollars and the regular appropriations of the Congress.

Incorporated under the laws of the Dominion of Canada as a non-profit company like its parent organization within the United States, Ducks Unlimited is guided by a board of directors composed of sportsmen, professional, and business men who serve without compensation. Some of the outstanding business leaders in the United States are represented in the managerial group of the association. DU began conservation work in Canada early in 1938 and in a decade had collected many hundreds of thousands of dollars from American sportsmen for field work across the border, for the basic purpose of assuring water under all conditions at times when it is needed the most on the Canadian waterfowl breeding areas.

The work of Ducks Unlimited has been concentrated largely in the southern region of the three provinces of Alberta, Saskatchewan, and Manitoba, once a veritable waterfowl paradise with its many lakes, ponds, marshes, sloughs, and potholes. But the advancement of agricultural interests and giant land drainage promotion schemes that followed World War I left sweeping acreages of land barren and useless for both man and waterfowl. The program of Ducks Unlimited has been to restore these areas by developments which will hold water in a permanent status and prevent the periodic drying up of the marshes with the changing seasons.

To quote from a recent DU publication:

These factors in the fight to maintain duck production at the highest possible levels along with the battle against adverse factors such as drouth, floods, fires, predators, and other destructive elements keeps DU on its toes the year around. Nature can be helpful but she can be destructive too. It is up to man to capitalize on her better moods and fight her when she becomes unkind. * * *

DU's work to increase waterfowl production and maintain the advantage already established is based upon developing in every conceivable way the most efficient use by man of land and water resources. DU projects are designed to improve domestic conditions and reduce losses of these important birds. * * *

Marsh restoration and management has become one of DU's most effective weapons. Examples of outstanding marsh stabilization by DU for managed production are the Bracken Dam (DU Project No. 50) and Knapp Dam (DU Project No. 63). These dams were built jointly by DU and the Manitoba government to improve wildlife production on 131,000 acres of Saskatchewan River Delta marsh. This area is being operated by the Manitoba Government as a wildlife production block. Development is designed to protect water levels against drought and flood. DU-raised birds, in average seasons, fly in prodigious numbers from this area.

An outstanding example of restoration is Big Grass Marsh (DU Project No. 1) in Manitoba. This marsh had been unwisely drained. Farm families trying to transform the muskeg into farms were starved out. Duck broods perished each year. Fires turned the peat into square miles of ash. Municipalities, government and private owners cooperated with DU to build dams to hold snow runoff waters and to restore lakes and ponds rather than to allow the ditches to dry away. Neighboring ranchers were organized into volunteer fire fighters. Fences were erected to bring grazing under control. Fire towers were built. Municipalities and private owners organized a cooperative to manage the marsh—under DU guidance, and to harvest the annual muskrat crop by which such areas are kept self-supporting.

The Big Grass Marsh development serves as a model for other western communities for greater duck production. It is a power of influence in checking Canadians to stop marshland drainage to the detriment of wildfowl.

Waterhen Marsh (DU Project No. 4), in central Saskatchewan, is another example of a marsh drained years ago for farmlands. Farming failed. DU, in cooperation with Saskatchewan authorities and citizens, built dams to restore the marsh.

In its basic program of putting water to work DU has developed a vast number of areas everywhere within the great duck nesting range that now serve water advantageously for breeding ducks. One of them is the Louisiana Lakes (DU Project No. 112-128) area in southern Alberta. It illustrates another type of development than Waterhen. Funds to build water controls were provided by hunters of Louisiana. Land and water were provided by authorities and ranchers. In this development, surplus irrigation water is led over the prairie to create 14 lakes and ponds for the ducks. Nesting population is dense and production heavy because of outstanding waterfowl food conditions and because of cooperation of Canadians with DU in the undertaking. Many others have been developed similarly.

Ducks Unlimited reports that more than 200 restoration projects have been completed, that thousands of acres of nesting territory have been improved, that predators such as crows have been reduced in numbers, that 50,000 ducks have been banded, that many miles of barbed wire fencing have been erected to protect nesting areas, that miles of fire guards have been cleared to prevent harmful marsh fires, and that DU has made the peoples of two nations more conservation-conscious.

All of these things are valuable in the international waterfowl management picture. May the restoration of marshes in Canada increase. May the improvement of habitat be carried forward to the fullest extent possible with the funds and manpower available. Certainly we in the Fish and Wildlife Service can testify to the fine work that Ducks Unlimited has done in spreading the gospel to American sportsmen that if wildfowling is to continue, the needs of the ducks must be considered first. DU has banded together into local chapters the most enthusiastic and ardent group of duck hunters that I have ever witnessed. The organization has done much to educate the youngsters as well as the oldsters of Canada that waterfowl represent an important resource and certain fundamentals of management must be observed if it is to be perpetuated.

To those in this great organization who have convinced themselves that the Ducks Unlimited program is the all-important part in waterfowl management, I recommend a more thorough investigation. To those scattered individuals who feel that they should be permitted to dictate seasons, bag limits, and management policies because they have contributed cash to the cause of waterfowl restoration, I would urge further study in order that they might understand more of the basic philosophies and principles that go into the management of a continental resource of the magnitude and value of waterfowl. To the other thousands of enthusiastic and sincere members of Ducks Unlimited I would plead for a continuation of the efforts to restore every acre of marsh, everywhere, that can be put back into waterfowl production. The need is as critical in Canada as anywhere on the North American continent while agricultural demands there are constantly pressing in upon the potholes and marshes that can provide the only kind of habitat which produces a continuing supply of ducks and geese.

National Wildlife Federation

Out of the first North American Wildlife Conference called in Washington, D. C., on February 5, 1936, by the late President Franklin D. Roosevelt, evolved a non-profit corporation known as the National Wildlife Federation. One of the proclaimed purposes of the conference was to focus attention upon the "many social and economic values that wildlife has for our people." The Federation was formed to disseminate pertinent facts, discoveries, and information relating to the preservation and restoration of wildlife through the conservation of soils, waters, plants, and forests.

Since its formation, the National Wildlife Federation has placed major emphasis upon needed programs for conservation education, not only among adult groups such as sportsmen's clubs, women's and garden clubs and youth organizations, but also in the public schools. One of the chief obstacles to the teaching of conservation in the public schools has been the lack of training on this subject among the teachers and in the courses provided in universities and teachers' colleges. The Federation has encouraged the establishment of conservation workshops giving brief courses for teachers during summer vacations.

The Federation has prepared a series of three handbooks edited by the distinguished educator, the late Dr. Henry Baldwin Ward, on the teaching of conservation. Many thousands of copies of these monographs have been placed with interested teachers.

The first edition of 250,000 copies of the conservation series "My Land and Your Land," consisting of four booklets, has been distributed to schools. They are fully illustrated in color and the text is written in an interesting story-book form. Louis Bromfield, author, farmer, and conservationist, wrote the foreword for each one of the booklets which cover grades three to eight in the public schools. These present in brief review the relationship of man to his environment and the need for conserving our natural resources.

J. N. "Ding" Darling, famous cartoonist, conservationist, former

Chief of the U. S. Biological Survey, and past president of the Federation, prepared an article called "Poverty or Conservation," several million copies of which have been distributed. This has made a profound impression upon our thinking concerning conservation needs. I recommend it highly.

"Botany and our Social Economy," prepared by Dr. Alexander C. Martin, of the Fish and Wildlife Service, is another Federation publication that points up our dependence upon the plants that nature so abundantly provides.

The publications which the Federation distributes show the vital importance of conserving through wise use and management the natural resources which have so bountifully been placed around us and which form the very basis of our existence as a nation.

Once each year the Federation issues a sheet of Wildlife Conservation Stamps which reproduce paintings by well-known nature artists. More than 450 subjects of animals, upland game, song and insectivorous birds, fishes, wild flowers and trees of the North American continent have been reproduced in natural colors. These stamps were first issued in 1938, and more than 10,000,000 sheets have since been distributed.

While strong emphasis is placed upon conservation education, the Federation renders significant service in other fields. It maintains a Servicing Division which distributes annually hundreds of thousands of leaflets to people who request conservation information. It maintains a Legislative Reference Service which lists and analyzes all of the bills introduced in the Congress having to do with any phase of conservation, such as agriculture, water pollution, fish and game, forests, irrigation and reclamation, parks and monuments, and public lands. Bills are followed through their course in the Congress and every action upon any of them is forwarded to several thousand affiliated groups—state game commisions, outdoor writers, garden clubs, conservation committees, and others throughout the country who manifest interest.

The Federation sponsors the annual National Wildlife Restoration Week, originally proclaimed by the late President Franklin D. Roosevelt. The week in which March 21st falls has been set aside by the Federation for Wildlife Week during which conservationists throughout the country conduct various programs from the platform, through the press, and over the radio calling to the attention of the American people the many values that are to be found in the out-of-doors and the need for protecting them.

The Federation has no dues but has affiliated groups in almost all of the states. It has a program of grants-in-aid for state affiliates, covering specific conservation projects and when approved by the Federation, money is set aside to carry them out.

The Federation sponsored the Pittman-Robertson Act, a similar program for aid to fisheries, and many other helpful long-range programs for the preservation of wildlife and the perpetuation of its pursuit.

This organization, with David A. Aylward of Boston as its president, and Carl D. Shoemaker, of Washington, as conservation director, has had a potent influence on wildlife conservation in the United States. There is no national legislation designed for the betterment of wildlife in America that is not influenced by the power of this volunteer and non-profit group. Carl Shoemaker for years served as secretary of the Senate Special Committee on the Protection of Wildlife Resources, and knows his way about the Capitol as does no other man in conservation. His influence can be found in every important conservation bill enacted in recent years, particularly those pertaining to waterfowl management.

Outdoor Writers Association of America

The outdoor writers, those journalists who write daily and weekly columns on the hunting, fishing, and outdoor news of the country, wield a potent influence on national conservation thinking. These prophets are supposed to predict exactly where the best fishing will be for next week end and how the ducks will fly on the very day the reader can get away from the office to do a little hunting. But the outdoor writers of the nation have a responsibility far beyond the mere enlightment of the public on how to take fish and game quicker, easier, and cheaper. They can and do constantly preach the fundamental principals of conservation—the restoration and management that must be observed if the sport is to be perpetuated.

The Outdoor Writers Association of America is an organization "whose members seek by their activities and writings, the conservation and preservation of our wildlife and other natural resources;

the teaching of better sportsmanship in all of our recreational journeyings into the outdoors, with dog and gun, or rod and reel." Although organized many years ago, it was not until 1939, when reorganized, that it became a really effective force in moulding public conservation opinion. At the time of the reorganization it had an enrollment of 37 members, of which only 20 had paid their dues. At the present time, the membership is well over 1,000.

Membership in the Outdoor Writers Association consists of columnists of the daily and weekly press, of radio commentators, and of some who have a regular schedule on television; of members of the staff and contributors to the national magazine field on outdoor subjects; of authors of outdoor books; of photographers and artists who specialize in the outdoor scene; and of those who work in education and public relations for organizations, agencies, and departments having to do with conservation on the state, Federal, and national organizational levels. Almost half of the membership consists of columnists for daily and weekly papers and for radio and television broadcasts.

Some 400 members have regular columns in the daily and weekly press, while a smaller group, less than 50, have syndicated outdoor columns which reach about 4,000 small daily and weekly newspapers. More than 175 members conduct regular programs on the radio and on television. Almost 300 members are affiliated with magazines as editors, or as staff members of departments of state and national outdoor organizations.

The Outdoor Writers Association has sponsored a unique little outdoor fraternity known as the "Brotherhood of the Jungle Cock." This was conceived at one of the annual meetings of the Association held at a state park at Catoctin, Maryland. The creed of the Order of the Jungle Cock has been spread throughout many states, and even in some foreign countries. Members in military activities during the war spread the Jungle Cock teachings to such far-away places as the islands of the South Pacific, to Japan, etc.

The creed of the Jungle Cock follows:

We who love angling, in order that it may enjoy practice and reward in the later generations, mutually move together towards a common goal—the conservation and restoration of American game fishes.

Towards this end we pledge that our creel limits shall always be less than the legal restrictions and always well within the bounty of Nature herself.

Enjoying, as we do, only a life estate in the out-of-doors, and morally charged in our time with the responsibility of handing it down unspoiled to tomorrow's inheritors, we individually undertake annually to take at least one boy a-fishing, instructing him, as best we know, in the responsibilities that are soon to be wholly his.

Holding that moral law transcends the legal statutes, always beyond the needs of any one man, and holding that example alone is the one certain teacher, we pledge always to conduct ourselves in such fashion on the stream as to make safe for others the heritage which is ours and theirs.

The Outdoor Writers as a group have always been on the side of sound conservation. The indefatigable president, J. Hammond Brown, of Baltimore, Maryland, regularly transmits current news from all parts of the continent to the membership. This group can be of increasing influence as the membership grows and as the national conservation movement brings together all interests and factions in a more united front.

Other Organizations

I would be remiss in chronicling the work of organizations if I did not acknowledge the fine efforts of the many other groups that cannot be given individual tribute in these pages: the Boone and Crocket Club, the Campfire Club, the American Forestry Association, the American Nature Association, the Wildlife Society, the American Society of Mammalogists, the Society of American Foresters, the Wilderness Society, the American Ornithologists' Union, and the numerous other national, state, and local organizations that have done much to further the cause of general wildlife conservation. Some have been interested in one phase and some in another, but all have been helpful and have contributed in this nation-wide program of restoring and protecting all wildlife. Waterfowl management has had its fair and just share of attention.

The national magazines that each month reach millions of people contribute materially to the better understanding of the wildlife problems of the nation. *Field and Stream, Sports Afield, Outdoor Life, Outdoors,* and *Outdoorsman* magazines deserve particular mention. The list is indeed revealing when one attempts to catalogue the groups of people, the organizations, and the varied interests that are attempting to do a better job of wildlife conservation. All can assist in taking to the public a better understanding of the basic problems involved in trying to perpetuate the fish and game resources that are desired by so many folks in a country that has no new frontiers. We must make our existing lands and waters more productive to meet the ever-expanding requirements of a growing national population. Such organizations of public spirited conservationists do much to spread that gospel.

Individuals and Clubs

A highly important contribution to the waterfowl management program is the work of the thousands of hunting clubs and of individual sportsmen. Many fine clubs and shooting preserves are maintained at great expense all along the Atlantic and Gulf coasts, where ducks and geese are given protection and food. Many excellent pools have been created, some by means of dikes and pumping stations, along the southern shore of Lake Erie. In fact, the famous Lake Erie marshes are almost entirely man-made. Similar developments have been made by wealthy sportsmen in Illinois, Arkansas, and Missouri, while in California gun clubs own and control thousands of acres of land on which water and food are provided. The famous McIlhenny preserve is a good example of what individuals can do to aid in waterfowl management. Although hunting is done on these acres, the membership is usually quite restricted and the total kill is light.

George T. Slade, at one time General Superintendent of the Great Northern Railroad and later of the Northern Pacific Railroad, who had spent many thousands of dollars in developing and improving his favorite shooting grounds near Dawson, North Dakota, donated the entire property to the Government at his death, and it is now maintained as the Slade National Wildlife Refuge in the status that he wished perpetuated—producing ducks for other sportsmen to enjoy.

Maurice L. Wertheim, a wealthy sportsman of New York City, recently provided in his will that upon his death his favorite shooting grounds "Steal-away," located along the Carmen River on the southern shore of Long Island, should pass to the Fish and Wildlife Service. This will be known as the Maurice L. Wertheim Migratory Waterfowl Refuge, to perpetuate his interest in continuing the resource for future generations. These fine sportsmen have amply demonstrated their interest in the future welfare of the ducks and geese.

(Author's note: If this idea is of interest to any readers in similar circumstances, my address is readily available.)

CHAPTER XIX

Your Waterfowl

WE HAVE come a long way in conservation thinking in this country since the days when the wild things of the plains, the forests, and the streams were considered as legitimate prey for the first who could reduce them to possession. The American public now has a much better understanding of the fundamental principles of wildlife protection, preservation, and management. The road has been rough and the progress slow, but year by year, rocks and boulders have been rolled from the trail as enlightened legislation and more generous funds have been made available for the perpetuation of wildlife. But the final chapter has not yet been written. Only Time will prove whether this American democracy of ours can insure for future generations what we and our forefathers have taken as God-given birthrights.

There are some who sincerely feel that hunting and fishing in the United States will eventually follow the pattern of that in England and other European countries. On the British Isles, good hunting and fishing have been maintained for hundreds of years in the midst of dense populations. But there, game and fish belong to the landlords. As such, those folks have a personal interest and they care for and protect wildlife as they do domestic stock. Game is not over-shot, even though surpluses are sold in the public markets. Fishing is never permitted to remove all of the fish from the streams.

But there, only the few are permitted to enjoy hunting and fishing. Here in the United States, we have attempted to manage the sport so that it will be available for all—rich and poor alike. In this country, where streams are overrun and crowded with fishermen who fight to see how many can surround a single pool without getting their flies and plugs tangled with those of their elbowing neighbors, the government agencies are supposed to supply enough fish so that each license buyer can get somewhere close to his creel limit every day. If that limit is set at less than 10, there is usually loud and long criticism of the Fish and Game Department. In England, on well-managed streams, two trout are often considered a reasonable limit. Few streams in the United States can naturally produce the size and quantity of finny trophies that the public expects.

Don't get me wrong! I would be the last to advocate the adoption of the Continental system of hunting and fishing instead of our own type of public participation in this most healthful of outdoor recreations. I am not yet ready to toss in the sponge. I cannot yet subscribe to the philosophy that the end is in sight; that our children, grandchildren, and great grandchildren will see no more of the kind of enjoyment that we of this generation have had. I think we can continue to maintain public hunting and fishing in the United States for many years—perhaps forever.

But if we do, the average individual is going to be forced to take a more realistic attitude than is shown by the bulk of the present-day license-buyers. He must be made to realize that wildlife is a product of the soil and water, and that it cannot be manufactured artificially and placed in the fields and streams for his special enjoyment on the day that he happens to have a chance to get away from his office. True, game-farm reared pheasants and quail can supplement the natural supply as can hatchery-raised trout, bass, and sunfish. But the amount that a man pays for his license lacks, by a long way, the cost of producing and liberating artifically-reared game and fish.

To Mr. John Q. Sportsman must come the realization that under our American system game and fish belong to the American public and they must be maintained not alone for the present but also for future generations. The taking of ducks, geese, quail, rabbits, trout, bass, deer, as well as all other species which are subjected to hunting and fishing laws, should be considered as representative of the American way of life. Pride should not rest in the man who is able to take the most—in the shortest length of time.

I would like to think that most sportsmen are high-minded, decent, law-abiding gentlemen, but many of the cases which come to the attention of administrative agencies certainly leave plenty of room for doubt. Perhaps we get more than our share of the trigger-happy boys who shoot without restraint and without any thought for their neighbors, their fellow-men, and the future of the sport. Perhaps we hear too many tales of the individuals who have no regard for the landowners on whose farms and ranches wildlife are produced and who deliberately leave gates open, who cut fences, who destroy property, and who even kill chickens, turkeys, and pigs to stow in the trunks of their automobiles to take home as trophies of the hunt. Such things are not rare occurrences. They happen daily, hourly, whenever there is an open season in any state of the Union.

I have been told at some of the public waterfowl meetings which we have conducted throughout the United States that the Fish and Wildlife Service personnel are anti-sportsman and anti-hunter in their thinking; that they consider nothing except absolute protection of the birds; that the Service would gladly see a complete closed season with no more hunting of any kind.

There is nothing farther from the truth than such assertions.

If we, at times, seem impatient it is because we continually receive a barrage of complaints and criticisms from those few individuals throughout the country who resent the restraints that prevent them from taking more than their share. I have had folks tell me that we should relax the regulations protecting waterfowl because we cannot enforce them; that people feel that they may violate the regulations with impunity because they do not agree with them. We have been told that the restrictions against baiting should be lifted because some folks continue to use bait as a means of taking waterfowl whenever and wherever they can get away with it. We have been told also that shooting hours are violated because some persons feel them too restrictive. The philosophy of a too-large majority of hunters is that the regulations should be made to suit their own ideas and their own conveniences rather than for the management of the resource. If that policy were followed for long there would soon be no need to issue protective regulations. There would be nothing left to protect.

The sportsmen of this county contribute substantially and materially to the wildlife management programs. In fact, their dollars finance almost the entire protective structure. State game licenses support the state fish and game departments since only in a few instances do the states contribute any substantial sum from general revenues. The tax on arms and ammunition provides the revenue for the Pittman-Robertson Federal Aid program which does so much toward the restoration of all forms of wildlife. The duck stamps provide much of the Federal funds for the waterfowl refuges, for enforcement, and for research. Many individuals maintain private clubs and sanctuaries and contribute to numerous organizations which seek to improve wildlife conditions. Sportsmen and license buyers have a great stake in this conservation program because they finance it. But money, by itself, is not enough.

More than 25 million people in these United States buy hunting and fishing licenses each year. True, many of these are duplicates, but true, also, many others do not buy licenses at all. In the latter group are those who like to take a chance on hunting or fishing without paying the usual fare. Also, there are youngsters who, by state and Federal laws, are not required to purchase licenses and duck stamps. In other cases, state statutes provide that land-owners need not buy licenses to hunt on their own properties. Thus it seems that we might take 25 million as the reasonable figure for those people who each year take to the woods and streams. It is utterly impossible for the small group of conservation officers in this country to make law-abiding citizens out of such a mass of hunters and anglers. Observance of the laws and regulations must come from within this group of sportsmen.

The state fish and game departments are now equipped with better personnel than at any time in the entire history of conservation. The Pittman-Robertson program has stabilized the technical staffs of the state administrations and the cooperative units have provided technicians that are doing a splendid job. The Fish and Wildlife Service has, as a heritage, a reputation for sincerity and honesty of purpose. It is equipped and staffed with the best people than can be employed. Your official agencies are attempting, with rare exceptions, to do a first-class job of managing the wildlife resource. Certainly they make mistakes and they will continue to do so, both in judgment and in public relations. Yet, how can a little handful of officials be expected to do all of the things that are necessary to provide adequate hunting and angling for one out of every five people in the United States?

How many hunters devote even one single day each year in attempting to improve game habitat in the areas where they hunt? How many go out to feed starving game in winter when snows are deep and cold is severe? How many try to do something for the landowner to repay him for the opportunity to hunt on his property? How many are willing to do something on their own for the production and protection of wildlife? How many of your acquaintances think about the needs of the birds instead of their own desires when they buy that license and take a little time off from a busy life to go afield?

Count 'em yourself! You know the people that you associate with—your friends on fishing and hunting expeditions. Compare those real conservationists with the numbers who are willing to arise in every rod and gun club meeting and shout to the high heavens that the state and Federal officials are complete and thorough failures and there should be a general overhaul of the official agencies. What we need more than anything else, if we are to preserve the free hunting and fishing heritage of this country, is a larger percentage of those who look first to the welfare of wildlife and secondly to their own desires.

The fate of *your* wildlife—of *your* outdoor recreation—depends upon *you* more than it does upon the elected and employed officials of the state and Federal conservation agencies. They can make rules and regulations designed to protect the resource—they can do their best to enforce them; but unless you 25 million hunters and fishermen in these United States cooperate and do something on your own, the case will eventually be hopelessly lost.

The chap who shoots more birds than his limit or who takes more fish than the regulations prescribe is not being smart and fooling the game warden. Instead, he is robbing his son and his son's son of one of the finest things that America has provided for its people. He is helping to bring to an end the success of a plan that has never survived in any other heavily populated country. He is contributing to the downfall of one of the greatest heritages that we have—one that folks all over the world envy and admire. He is not cheating that "nosey game warden"—the wildlife does not belong to him. Rather the man who violates the game laws is cheating those millions of young Dickies and Johnnys and Joes and Georges and Dorises and Marys and Myrtles who would also like some day to take part in that great American experiment.

"FOR FISH AND BIRDS"

For fish and birds I make this plea
May they be here long after me;
May those who follow hear the call
Of old Bobwhite in Spring and Fall;
And may they share the joy that's mine
When there's a trout upon the line.
I found the world a wondrous place,
A cold wind blowing in my face
Has brought the wild ducks in from sea;
God grant the day shall never be
When youth upon November's shore
Shall see the mallards come no more!
Too barren was the earth for words
If gone were all the fish and birds.
Fancy an age that sees no more
The mallards winging in to shore;
Fancy youth with all its dreams
That finds no fish within the streams.
Our world with life is wondrous fair;
God grant we do not strip it bare!

—Edgar A. Guest

Bibliography

Anonymous
    1931. More Waterfowl by Assisting Nature. 106 P., Illus. More Game Birds in America, 500 Fifth Avenue, New York, August, 1931.

Barbour, Frederick K.
    1947. Duck Shooting along the Atlantic Tidewater, Wm. Morrow & Co., New York, N. Y.

Bennett, Logan J.
    1938. The Blue-winged Teal; its ecology and management, 144 P., Illus. Collegiate Press, Inc., Ames, Iowa.

Cameron-Jenks.
    1929. The Bureau of Biological Survey, Its history, activities and organizations. Johns Hopkins Press, Baltimore, Maryland.

Camp, Raymond R., Editor
    The Hunter's Encyclopedia, 1152 P., Illus. Stackpole and Heck, Inc., Harrisburg, Pa., and New York, N. Y.

Cartwright, B. W.
    1940. Restoration of Waterfowl Habitat in Western Canada. P. 377. Transactions of the Fifth North American Wildlife Conference, Washington, D. C.

Kenney, F. R. and McAtee, W. L.
    1939. The Problem: Drained Areas and Wildlife Habitats, Yearbook Seperate No. 1611. U. S. Department of Agriculture.

Colvin, Stephen Sheldon and Bagley, Wm. Chandler.
    1915. Human Behavior. The Macmillan Company, N. Y.

Connett, Eugene V.
- 1947. Duck Shooting Along the Atlantic Tidewater. Wm. Morrow & Co., New York, N. Y.

Cottam, Dr. Clarence
- 1939. Food Habits of North American Diving Ducks. Tech. Bul. No. 643. U. S. Department of Agriculture, Washington, D. C.

Crowell, A. Elmer
- 1947. Duck Shooting along the Atlantic Tidewater. Wm. Morrow & Co., New York.

Day, Albert M.
- 1946. June. Goose Business. *Outdoorsman,* Chicago, Ill.
   October. What's Happened to our Ducks. *Outdoors,* Boston, Mass.

Gabrielson, Dr. Ira N.
- 1941. Wildlife Conservation. 250 P. Illus. Macmillan Co., New York.
- 1943. Wildlife Refuges. 245 P. Illus. Macmillan Co., New York.

Hornaday, Wm. T.
- 1912. Our Vanishing Wildlife, its extermination and preservation. 411 P. Illus. Chas. Scribner's Sons, New York.
- 1913. Our Vanishing Wildlife. New York Zoological Soc., New York.

Kalmbach, E. R.
- 1937. Crow-Waterfowl Relationships. Based on preliminary Studies on Canadian Breeding Grounds. Circ. 433, U. S. Dept. of Agriculture.
- 1940. Bird Control: A Statement of Federal Policies with a Suggested Method of Approach. P. 195. Transactions of the Fifth North American Wildlife Conference, Washington, D. C.

Kortright, Francis H.
- 1942. The Ducks, Geese and Swans of North America. American Wildlife Institute, Washington, D. C.

Lewis, Harrison E.
- 1946. Management of Canada's Wildlife Resources. Eleventh North American Wildlife Conference.
- 1947. Wildlife Conditions in Canada. Twelfth North American Wildlife Conference, Wildlife Management Institute, Washington, D. C.

McIlhenny, E. A.
- 1934. Bird City. The Christopher Publishing House, Boston, Mass.
   Martin, Dr. A. C. and F. M. Uhler.
- 1939. Food of Game Ducks in the United States and Canada. Tech. Bul. 634., U. S. Department of Agriculture, Washington, D. C.

Miner, Jack.
- 1923. Jack Miner and the Birds. The Reilly & Lee Co., Chicago, Ill.

Parks, Richard L.
 1947. Duck Shooting Along the Atlantic Tidewater. Wm. Morrow & Co., New York, N. Y.

Person, H. S.
 1935. Little Waters, A Study of headwater streams and their little waters, their use and relations to the land. 82 P. Illus. U. S. Government Printing Office, Washington, D. C., 1936.

Pearson, T. Gilbert
 1937. Adventures in Bird Protection. D. Appleton-Century Co., Inc., New York.

Phillips, John C.
 1929. Shooting Stands of Eastern Massachusetts. Riverside Press, Cambridge, Mass.

Pirnie, Miles D.
 1938. Restocking of the Canada Goose successful in southern Michigan. Third North American Wildlife Conference, Washington, D. C.
 1941. Muskrats in the Duck Marsh. Sixth North American Wildlife Conference. Washington, D. C.

Quortrup, E. R. and Sudheimer, R. L.
 1942. Research Notes on Botulism in Western Marsh Areas with Recommendations for Control. P. 284. Transactions of the Seventh North American Wildlife Conference, Washington, D. C.

Salyer, J. Clark.
 1936. Practical Waterfowl Management. North American Wildlife Conference.
 1941. Are We Using the Refuge Idea Wisely? Thirty-fifth Conv. International Assoc. Game Fish and Conservation Commissioners.

Titus, Harold
 1934. The Truth about the Izaak Walton League. *Field and Stream* (Jan. 1943), New York.

Velie, Lester.
 1948. What Are We Going to Do for Water? *Collier's* (May 15, 1948), Crowell-Collier Pub. Co., New York.

Wetmore, Alexander
 1919. Lead Poisoning in Waterfowl. Bul. 793, U. S. Dept. of Agriculture.

Wheeler, Chas. E.
 1947. Duck Shooting along the Atlantic Tidewater. Wm. Morrow & Co., New York.

Index

A.

American Bird Banding Association formed, 58
American brant, 76
American Ornithologists' Union, 34
American Society of Bird Restorers, 36
Antelope, viii, x, 5, 162
Arkansas National Wildlife Refuge, 174
Arctic tern, 64 (migration route)
Arrowwood Refuge, 189, 196
Atlantic Flyway, 76, 87, 108 (hunting regulations), 172, 187
Audubon Society, 43
Australian wild rabbit, 35

B.

Bacillus botulinus, 206
Baldpate, 86, 78
Banding, 59
Banding organizations; 59
Banding results, 65
Band-tailed pigeon, 48
Bartsch, Paul, 58
Battery, 12
Bear, 192
Bear River National Wildlife Refuge, 207
Beaver, 138, 162, 192
Beaver in waterfowl management, 188
Bird-banding, 58
Bird migration, 55
Blackbeard Island Refuge, 196
Blackbirds, 210
Black brant, 84 (flyway)
Black-crowned night heron, 58
Black duck, 8, 9, 10, 74, 75 (flyway), 76, 79, 87, 172, 188, 258
Blackwater Refuge, 187
Blind, bush, 12
  staked, 12
Blinds, 140
Bluebill, 76, 83 (flyway)
Blue goose, 81 (flyway), 185, 196
Blue-winged teal, 44, 64, 75 (flyway), 76, 83 (flyway), 277
Blue-wings, 158
Boone and Crocket Club, 43
Botulism, 64, 206, 231
Brant, 46, 172
  American, 76
  black, 84 (flyway)
Breeding colonies established, 195
Breeding refuges, 157
Broadbill, 11
Buffalo, 34 (closed season), 256
Bull snake, 189
Bush blind, 12

C.

Cackling goose, 84 (flyway), 86
Canada goose, 14, 37, 44, 45, 67 (migration), 75, 76, 79, 83 (flyway), 84, 86, 87, 162 (nesting), 172, 173, 185, 187, 188, 195, 196, 233 (nesting), 235, 256
Canvasback, 14, 45, 74, 76, 86, 157
Census sampling, 102
Central Flyway, 76, 81, 87, 109 (hunting regulations)
Chautauqua National Wildlife Refuge, 80
Chimney swift, 63 (wintering ground)
Cinnamon teal, 83 (flyway), 273
Conservation in Canada, 256
 private, 259
Conservation organizations, 289
Controlled burning, 186
Cooperative units, 204
Coots, 59 (banding), 94, 267, 271, 276
Coyote, x, 189, 208
Crab Orchard Refuge, 196
Crane, Florida, 175
 little brown, 48, 86, 176
 sandhill, 44, 48, 137, 162 (breeding), 176
 whooping, 48, 174
Crescent Lake National Wildlife Refuge, 185
Cripple losses, 213
Crow, 189, 208
Crying bird, 175
Curlew, 48
 long-billed, 176, 185

D.

Decoys, 8, 9, 140
Deer, 5, 192, 193 (transplanting), 195, 196 (hunting)
 mule, 162
Depredations, 210
Depression, effects of, 224
Des Lacs Refuge, 158, 160
Duck, baldpate, 86, 87
 black, 8, 9, 10, 74, 75 (flyway), 76, 79, 87, 172, 258
 bluebill, 76, 83 (flyway)
 broadbill, 11
 canvasback, 45, 74, 76, 86, 157
 eider, 48
 goldeneye, 86
 lesser scaup, 65, 75 (flyway), 80 (flyway), 277

 mallard, 10, 80 (flyway), 81 (slaughter), 84, 87, 96, 157, 158, 172, 188, 189
 pintail, 64, 75 (flyway), 81 (flyway), 83 (flyway), 84 (migration route), 86, 87, 96, 157, 172, 189, 276, 277
 redhead, 9, 10, 45, 74, 75 (flyway), 76, 83 (flyway), 84
 ring-neck, 87, 188
 ruddy, 9, 10, 14, 45
 scaup, 9, 10, 74, 86, 87, 157, 172
 spoonbill, 84 (migration route)
 wood, 44, 48, 80 (flyway)
Duck blind, 9
Duck stamp, 109
Duck stamp law, 151
Duck trapping, 126
Duck traps, 128
Ducks Unlimited, 300

E.

Eider duck, 48
Elk, 5, 33, 195 (transplanting)
English sparrow, 34, 35, 37
European house sparrow, 36
Everglades National Wildlife Refuge, 176
Exotic species, 36

F.

Farming, 186 (management)
Federal agents, 117
Federal Bird Sanctuary, first in Canada, 256
Federal conservation machinery, 252
Federal Refuge, 148
Federal-State relationship, 223
Field trials, 196
Florida crane, 175
Flying fox, 35
Flyway, Atlantic, 76, 87, 108 (hunting regulations), 172, 187
 Central, 76, 81, 87, 109 (hunting regulations)
 Mississippi, 76, 87, 108 (hunting regulations)
 Pacific, 76, 87, 109 (hunting regulations)
Flyway biologists, 94
Flyways, 71
Food habits, 210
Forestal y de Caza, 269
Fox, 15

G.

Gadwall, 83 (flyway), 84 (migration route), 158

INDEX

Game management agents, 117 (requirements)
Game wardens, 117
Golden-eye, 86
Goose, 167, 251
  blue, 81 (flyway), 185, 196
  cackling, 84 (flyway), 86
  Canada, viii, x, 14, 44, 45, 75, 76, 79, 83 (flyway), 84, 86, 87, 162 (nesting), 172, 173, 185, 187, 188, 195, 196, 233 (nesting), 235, 256
  Ross's, 175
  snow, 96, 172, 185, 196
  whie-fronted, 76, 79, 83 (flyway), 86, 96
Grazing, 185
Great blue heron, 65 (migration)
Green-headed mallard, 277
Green-winged teal, 83 (flyway), 84 (migration route), 96, 277
Grouse, 186, 196 (hunting)
  ruffed, 45, 46
  sage, 162
Gulls, 119

H.

Hawks, 15
Heron, great blue, 65 (migration)
Hungarian partridge, 258
Hunters' take, 100

I.

Insecticides, 213
International cooperation, 260
Inventory, winter, 99
Izaak Walton League of America, 296

L.

Lacey Act, 41, 138
Lead poisoning, 206
League of American Sportsmen, 43
Lesser scaup, 65, 75 (flyway), 80 (flyway), 277
Limpkin, 175
Little brown crane, 48, 86, 176
Long-billed curlew, 176, 185
Lower Klamath Refuge, 187
Lower Souris Refuge, 158
Lynx, 15

M.

Magpie, 189
Malheur National Wildlife Refuge, 161, 175
Mallard, x, 10, 65 (longevity), 80 (flyway), 81 (slaughter), 83 (flyway), 84, 87, 96, 157, 172, 188, 189
  greenhead, 277

Management by flyways, 87
Market hunters, 44
Market hunting, 139
Mattamuskeet National Wildlife Refuge, 171
Medicine Lake Refuge, 189
Mexican waterfowl situation, 263
Migration, 34
Migratory Bird Convention Act, 50, 251
Migratory bird hunting stamp sales, 231
Migratory bird sanctuaries, 254
Migratory Bird Treaty Act, 43, 47, 48, 120 (violations), 135 (violation), 137
Miner, Jack, 61
Mink, 15, 236
Mississippi Flyway, 76, 79, 87, 108 (hunting regulations)
Missouri vs. Holland, 50
Mongoose, 35
Moose, 15, 192
Mud Lake Refuge, 189
Mule deer, 162
Muskrat, 195 (transplanting), 235, 236, 243
Muskrats in waterfowl management, 188

N.

National Audubon Society, 289
National Elk Refuge, 175
National Waterfowl Committee, 111
National Wildlife Federation, 305
Necedah Refuge, 196
Nesting ground conditions, 96
Nuttall Ornithological Club, 34

O.

Okefenokee National Wildlife Refuge, 175
Old squaw, 86
Outdoor Writers Association of America, 307
Owls, 15

P.

Pacific Flyway, 76, 84, 87, 109 (hunting regulations)
Parker River Refuge, 189
Partridge, Hungarian, 258
Passenger pigeon, 33
Pelican, white, 176
Pelican Island Sanctuary, 148
Pest plant control, 212
Pheasant, 167, 195 (transplanting)
Pigeon, 46
  band-tailed, 48

Pintail, 64, 75, 81 (flyway), 83 (flyway), 84 (migration route), 86, 87, 96, 157, 172, 189, 276, 277
Pittman-Robertson Federal Aid in Wildlife Restoration Act, 229
Plover, 46
Pocahontas Fowling Club records, 15
Predation, 208
Predator control on wildlife refuges, 189
Protective legislation, 39, 41
Public hunting, 196

Q.

Quail, 45, 196 (hunting)

R.

Rabbit, Australian wild, 35
Raccoon, 193 (transplanting), 213
Rail, 46
Rat, 35
Recreation, 189
Redhead, 9, 10, 45, 74, 75 (flyway), 76, 83 (flyway), 84
Red Rock Lakes Refuge, 175
Refuge, Arrowwood, 189, 196
  Blackbeard Island, 196
  Blackwater, 187.
  Crab Orchard, 196
  Crescent Lake National Wildlife, 185
  Des Lacs, 158, 160
  Everglades National Wildlife, 176
  Lower Klamath, 187
  Lower Souris, 158
  Malheur National Wildlife, 161, 175
  Mattamuskeet National Wildlife, 171
  Medicine Lake, 189
  Mud Lake, 189
  National Elk, 175
  Necedah, 196
  Okefenokee National Wildlife, 175
  Parker River, 189
  Red Rock Lakes, 175
  Ruby Lakes National Wildlife, 175
  Sabine National Wildlife, 185
  Sacramento National Wildlife, 167, 175
  Saint Marks Migratory Bird, 80, 175
  Salt Plains, 196
  Seney National Wildlife, 176, 195
  Souris National Wildlife, 195
  Tennessee National Wildlife, 237
  Tule Lake, 187
  Upper Souris, 158
  Valentine National Wildlife, 185
  Wheeler, 187
  White River, 196

  Wichita Mountains Wildlife, 193
Refuge development, 181
Refuge management, 181
Refuge system, 153
Refuges, 155
  breeding, 157
  waterfowl, 155
Regulations, 107
  how made, 111
Research into management problems, 201
Ringneck, 87, 188
Ruby Lakes National Wildlife Refuge, 175
Ruddy duck, 9, 10, 14, 45
Ruffed grouse, 45, 46

S.

Sabine National Wildlife Refuge, 185
Sacramento National Wildlife Refuge, 167, 175
Sage grouse, 162
Saint Marks Migratory Bird Refuge, 80, 175
Salt Plains Refuge, 196
Sanctuaries, 145, 255
Sanctuary, Federal bird, 256
Sandhill crane, 44, 48, 137, 162 (breeding), 176
Scaup, 9, 10, 74, 86, 87, 157, 172
Seney National Wildlife Refuge, 176, 195
Shoveller, 81 (flyway), 84, 158, 189
Skunk, 10, 189, 213
Snakes, 15
Snipe, 46
Snow goose, 96, 172, 185, 196
Souris National Wildlife Refuge, 195
Sparrow, English, 37
Spoonbill, 84 (migration route)
Sprig, 158
Staked blind, 12
State activities, 219
State interest in waterfowl, 222
Swan, 45, 46, 48, 86, 96 (nesting), 172, 251
  trumpeter, 175 (range)
  whistling, 83

T.

Teal, 188
  blue-winged, 44, 64, 75 (flyway), 76, 83 (flyway), 158, 277
  cinnamon, 83 (flyway), 273
  green-winged, 83 (flyway), 84 (migration route), 96, 277
Tennessee National Wildlife Refuge, 237

INDEX

Timber management, 186
Transact sampling, 95
Transplanting, 193
Trapping for transplanting, 193
Trumpeter swan, 175 (range)
Tule Lake Refuge, 187
Turtles, 189

U.

Upland game, 191
Upper Souris Refuge, 158

V.

Valentine National Wildlife Refuge, 185

W.

Waterfowl conservation in Canada, 247
Waterfowl investigations, 243
Waterfowl management, 91
Waterfowl refuges, 155
Waterfowl restoration program, 150
Waterfowl surveys, 243
Weeks-McLean Law, 46
Wheeler Refuge, 187

Whistling swan, 83
White pelican, 176
White-fronted goose, 76, 79, 83 (flyway) 86, 96
White River Refuge, 196
Whooping crane, 48, 174
Wichita Mountains Wildlife Refuge, 193
Widgeon, 81 (flyway), 84 (migration route), 158, 277
Wildcat, 15
Wild geese, 52
Wildlife G-men, 117
Wildlife Management Institute, 292
Wildlife refuges, 155
Wild turkey, 33, 193 (transplanting)
Winter inventory, 99
Wood Buffalo Park, 256
Woodcock, 46, 186
Wood duck, 44, 48, 80 (flyway)

Y.

Your waterfowl, 315